SpringerBriefs in Statistics

JSS Research Series in Statistics

Editors-in-Chief

Naoto Kunitomo
Akimichi Takemura

Series editors

Genshiro Kitagawa
Tomoyuki Higuchi
Yutaka Kano
Toshimitsu Hamasaki
Shigeyuki Matsui
Manabu Iwasaki
Yasuhiro Omori
Masafumi Akahira

The current research of statistics in Japan has expanded in several directions in line with recent trends in academic activities in the area of statistics and statistical sciences over the globe. The core of these research activities in statistics in Japan has been the Japan Statistical Society (JSS). This society, the oldest and largest academic organization for statistics in Japan, was founded in 1931 by a handful of pioneer statisticians and economists and now has a history of about 80 years. Many distinguished scholars have been members, including the influential statistician Hirotugu Akaike, who was a past president of JSS, and the notable mathematician Kiyosi Itô, who was an earlier member of the Institute of Statistical Mathematics (ISM), which has been a closely related organization since the establishment of ISM. The society has two academic journals: the Journal of the Japan Statistical Society (English Series) and the Journal of the Japan Statistical Society (Japanese Series). The membership of JSS consists of researchers, teachers, and professional statisticians in many different fields including mathematics, statistics, engineering, medical sciences, government statistics, economics, business, psychology, education, and many other natural, biological, and social sciences.

The JSS Series of Statistics aims to publish recent results of current research activities in the areas of statistics and statistical sciences in Japan that otherwise would not be available in English; they are complementary to the two JSS academic journals, both English and Japanese. Because the scope of a research paper in academic journals inevitably has become narrowly focused and condensed in recent years, this series is intended to fill the gap between academic research activities and the form of a single academic paper.

The series will be of great interest to a wide audience of researchers, teachers, professional statisticians, and graduate students in many countries who are interested in statistics and statistical sciences, in statistical theory, and in various areas of statistical applications.

More information about this series at http://www.springer.com/series/13497

Naoto Kunitomo · Seisho Sato
Daisuke Kurisu

Separating Information Maximum Likelihood Method for High-Frequency Financial Data

 Springer

Naoto Kunitomo
School of Political Science and Economics
Meiji University
Tokyo
Japan

Daisuke Kurisu
School of Engeneering
Tokyo Institute of Technology
Tokyo
Japan

Seisho Sato
Graduate School of Economics
The University of Tokyo
Bunkyo-ku, Tokyo
Japan

ISSN 2191-544X   ISSN 2191-5458 (electronic)
SpringerBriefs in Statistics
ISSN 2364-0057   ISSN 2364-0065 (electronic)
JSS Research Series in Statistics
ISBN 978-4-431-55928-3   ISBN 978-4-431-55930-6 (eBook)
https://doi.org/10.1007/978-4-431-55930-6

Library of Congress Control Number: 2018940416

Printed on acid-free paper

This Springer imprint is published by the registered company Springer Japan KK part of Springer Nature
The registered company address is: Shiroyama Trust Tower, 4-3-1 Toranomon, Minato-ku, Tokyo 105-6005, Japan

Preface

In the last decade, considerable interest has been paid to the problem of estimating integrated volatility using financial high-frequency data. It is now possible to use copious high-frequency data including stock markets and foreign exchange markets. Although the statistical literature contains some discussion on estimating continuous stochastic processes, earlier studies often ignored the presence of micro-market noise when trying to estimate the volatility of the underlying stochastic processes. Because the micro-market noise is important variously to analyze high-frequency financial data both in economic theory and in statistical measurement, several new statistical estimation methods have been developed. The main purpose of this book is to develop a new statistical approach, which is called the separating information maximum likelihood (SIML) method, for estimating integrated volatility and integrated covariance by using high-frequency data in the presence of possible micro-market noise.

In April, 2007, I (Naoto Kunitomo) was invited to Osaka University as the first visiting Osaka Stock Exchange (OSE) Professor by Kazuhiko Nishina and Hideo Nagai. Back then, I assumed that I knew the basics about stochastic analysis and financial economics, the former having been invented by the Japanese mathematician Kiyoshi Itô and refined by subsequent mathematicians at Osaka and Kyoto. I had co-authored a book in Japanese on mathematical finance and financial derivatives (Kunitomo and Takahashi 2003), which was awarded as the 2003 Nikkei Prize. I asked some staff of OSE for access to their high-frequency data because the Nikkei-225 Futures index, one of OSE's major financial products, had been successful as the major financial derivatives actively traded in Japan since 1987.[1] Upon examining their data, I realized that my understanding on both stochastic analysis

---

[1]It is known in finance that Dôjima rice market at Osaka in the seventeenth century was the oldest organized futures market. The trade of modern Nikkei-225 Futures was started in 1987 as the first financial futures in Japan. Osaka Stock Exchange (OSE) and Tokyo Stock Exchange (TSE) were merged as Japan Exchange Group (JPX) in 2013.

and real financial data was too poor to allow meaningful analysis, and I decided to start investigating the problem of financial high-frequency data further.

We inclined to think that we have more accurate information on hidden parameter if we have finer observations. Apparently, it is not the case for the estimation problem of volatility and co-volatilities of financial prices when we use high-frequency financial data. After a while, it was fortunate for me to have a new idea to measure volatilities and co-volatilities from high-frequency financial data and began to write the first paper on the method that we termed the *separating information maximum likelihood* (SIML) method. I asked Seisho Sato, then at the Institute of Statistical Mathematics and now at the University of Tokyo, to do some simulations and data analysis, whereupon he kindly joined my research project. Since then, Sato and I have tried to investigate statistical estimation problem on the volatility and covariation by using high-frequency data. Our approach seems to be novel, and there have been several applications of the basic SIML method. At the last moment, Daisuke Kurisu, who is a graduate student at the University of Tokyo, joined our research project because he was interested in jumps in financial markets and the application of stochastic analysis of jumps. I would also like to mention Hiroumi Misaki, then a graduate student and now at Tsukuba University, who conducted some simulations in the early stage. Because 10 years have passed since we started the research project, it is a good time for us to summarize our research activities and results in a coherent way.

This book is a summary of our joint research project on the SIML method for financial applications. We hope that it gives many readers a better understanding of the high-frequency financial data problem and also that it will be a good starting point to investigate the unsolved related topics in future.

During the process of preparing manuscripts, we have received several comments from researchers including Hiroumi Misaki, Wataru Ôta, Yoshihiro Yajima, and Katsumi Shimotsu, and we thank them for their comments to parts of the earlier versions. This work was supported by JSPS Grant-in-Aid for Scientific Research No. 25245033 and 17H02513.

Tokyo, Japan   Naoto Kunitomo
December 2017

Reference

Kunitomo, N., and A. Takahashi. (2003). *A Foundation of Mathematical Finance: An Application of Malliavin Calculus and Asymptotic Expansion*. Toyo-Keizai (in Japanese).

Contents

Chapter 1
Introduction

Abstract We introduce recent issues and research around volatility estimation based on high-frequency financial data. Previous studies often ignored the presence of micro-market noises, thereby obtaining misleading estimation results. In this book, we propose the separating information maximum likelihood (SIML) method.

Recently in the field of financial econometrics, considerable interest has been paid to the problem of estimating integrated volatility using high-frequency financial data. It is now possible to use copious high-frequency data in financial markets including stock markets and foreign exchange rates markets. Although the statistical literature contains some discussion on estimating continuous-time stochastic processes, earlier studies often ignored the presence of micro-market noise in financial markets when trying to estimate the volatility of the underlying stochastic process. Because micro-market noise is important variously to analyze high-frequency financial data both in economic theory and in statistical measurement, several new statistical estimation methods have been developed. For further discussion on the related topics, see Zhou (1998), Andersen et al. (2000), Gloter and Jacod (2001), Ait-Sahalia et al. (2005), Hayashi and Yoshida (2005), Zhang et al. (2005), Ubukata and Oya (2009), Barndorff-Nielsen et al. (2008), Christensen et al. (2009), Ait-Sahalia and Jacod (2014), and Camponovo et al. (2017) among others.

The main purpose of this book is to develop a new statistical method for estimating integrated volatility and integrated covariance by using high-frequency data in the presence of possible micro-market noise. Kunitomo and Sato (2008a, unpublished) originally proposed the separating information maximum likelihood (SIML) method. Subsequently, Kunitomo and Sato (2008b, 2010, 2011, 2013) investigated some further properties of the SIML method. The SIML estimator of integrated volatility and covariance for the underlying continuous (diffusion-type) process is represented as a specific quadratic form of returns. As we show in Chap. 3, the SIML estimator has reasonable asymptotic properties: It is consistent and asymptotically normal when the sample size is large and the integrated volatility is time-changing under general situations including some non-Gaussian processes and volatility models. When the integrated volatility is stochastic, we require the concept of stable convergence and there is a further technical problem involved, see Jacod and Shiryaev (2003), Jacod

© The Author(s) 2018
N. Kunitomo et al., *Separating Information Maximum Likelihood Method for High-Frequency Financial Data*, JSS Research Series in Statistics, https://doi.org/10.1007/978-4-431-55930-6_1

and Protter (2011). Gloter and Jacod (2001) developed the maximum likelihood (ML) estimation of a one-dimensional diffusion process with measurement errors and our method could be interpreted as a modification of their procedure. There have been related studies by Zhou (1998) and Barndorff-Nielsen et al. (2008). However, the SIML approach differs from their methods in certain aspects as well as embodying novel features.

The main motivation for our study is that it is usually difficult to handle the exact likelihood function and more importantly the ML estimation lacks robustness if the assumption of a Gaussian distribution does not hold in the underlying multivariate continuous stochastic process with micro-market noise. This aspect is quite important for analyzing multivariate high-frequency data in stock markets and the associated futures markets. Instead of calculating the full likelihood function, we try to separate the information on the signal and noise from the likelihood function and then use each type of information separately. This procedure simplifies the maximization of the likelihood function and makes the estimation procedure applicable to multivariate high-frequency data. We call our estimation method the separating information maximum likelihood (SIML) estimator because it represents an interesting extension of the standard ML estimation method. The main merit of SIML estimation is its simplicity and robustness against possible mis-specifications of the underlying stochastic processes. Then, it can be extended in several directions and can be used practically for multivariate (high-frequency) financial time series. Not only does the SIML estimator have desirable asymptotic properties in situations including some non-Gaussian processes and volatility models, it also has reasonable finite-sample properties. The SIML estimator is asymptotically robust in the sense that it is consistent when the noise terms are weakly dependent and endogenously correlated with the efficient market price process. The SIML method was developed in an empirical study of the multivariate high-frequency Nikkei-225 Futures data for risk hedging problem, and we noticed that in real applications, we must consider the micro-market structure and noise as illustrated in Chap. 4.

The plan of this book is as follows. In Chap. 2, we discuss on the underlying background to volatility and high-frequency econometrics. In Chap. 3, we introduce the basic model and the SIML estimation of integrated volatility and integrated covariances with micro-market noise as well as showing the asymptotic properties of the SIML estimator. In Chap. 4, we discuss the finite-sample properties of the SIML method and an application to the Nikkei-225 Futures data at the Osaka Securities Exchange (OSE). In Chap. 5, we detail the mathematical derivations of the asymptotic results given in Chap. 3. In Chaps. 6 and 7, we investigate the asymptotic robustness of SIML estimation when we have the round-off errors and price adjustment mechanisms. In Chap. 8, we propose the local SIML method, which is an extension of the basic SIML method introduced in Chap. 3. In Chap. 9, we consider the estimation of the quadratic variation of Ito's semi-martingales including jumps. Finally, in Chap. 10, we make some final comments and suggest possible extensions of the SIML method.

References

Ait-Sahalia, Y., and J. Jacod. 2014. *High-frequency financial econometrics*. University Press.

Ait-Sahalia, Y., P. Mykland, and L. Zhang. 2005. How often to sample a continuous-time process in the presence of market microstructure noise. *The Review of Financial Studies* 18–2: 351–416.

Andersen, T.G., T. Bollerslev, F.K. Diebold, and P. Labys. 2000. The distribution of realized volatility. *Journal of the American Statistical Association* 96: 42–55.

Barndorff-Nielsen, O., P. Hansen, A. Lunde, and N. Shephard. 2008. Designing realized kernels to measure the ex-post variation of equity prices in the presence of noise. *Econometrica* 76–6: 1481–1536.

Camponovo, L., Y. Matsushita, and T. Otsu. 2017. *Empirical likelihood for high frequency data*. Unpublished Manuscript.

Christensen, Kinnebrock, and Podolskij. 2009. Pre-averaging estimators of the ex-post covariance matrix in noisy diffusion models with non-synchronous data. Unpublished Manuscript.

Gloter, A., and J. Jacod. 2001. Diffusions with measurement errors. II : optimal estimators. *ESAIM: Probability and Statistics* 5: 243–260.

Hayashi, T., and N. Yoshida. 2005. On covariance estimation of non-synchronous observed diffusion processes. *Bernoulli* 11 (2): 359–379.

Jacod, J., and A.N. Shiryaev. 2003. *Limit theorems for stochastic processes*. Berlin: Springer.

Kunitomo, N. and S. Sato. 2008a. Separating information maximum likelihood estimation of realized volatility and covariance with micro-market noise. Discussion Paper CIRJE-F-581, Graduate School of Economics, University of Tokyo. http://www.e.u-tokyo.ac.jp/cirje/research/dp/2008.

Kunitomo, N. and S. Sato. 2008b. Realized Volatility, Covariance and Hedging Coefficient of Nikkei-225 Futures with Micro-Market Noise. Discussion Paper CIRJE-F-601, Graduate School of Economics, University of Tokyo.

Kunitomo, N. and S. Sato. 2010. Robustness of the separating information maximum likelihood estimation of realized volatility with micro-market noise. CIRJE Discussion Paper F-733, University of Tokyo.

Kunitomo, N., and S. Sato. 2011. The SIML estimation of realized volatility of Nikkei-225 futures and hedging coefficient with micro-market noise. *Mathematics and Computers in Simulation* 81: 1272–1289.

Kunitomo, N., and S. Sato. 2013. Separating information maximum likelihood estimation of realized volatility and covariance with micro-market noise. *North American Journal of Economics and Finance* 26: 282–309.

Ubukata, M., and K. Oya. 2009. A test for dependence and covariance estimator of market microstructure noise. *Journal of Financial Econometrics* 7 (2): 106–151.

Zhang, L., P. Mykland, and Y. Ait-Sahalia. 2005. A tale of two time scales: determining integrated volatility with noisy high-frequency data. *Journal of the American Statistical Association* 100: 1394–1411.

Zhou, B. 1998. F-consistency, De-volatilization and normalization of high frequency financial data. In *Nonlinear modeling of high frequency time series*, ed. C. Dunis, and B. Zhou, 109–123. New York: Wiley.

Chapter 2
Continuous-Time Models and Discrete Observations for Financial Data

Abstract We introduce continuous-time financial models and the stochastic processes of diffusions and jumps. This chapter reviews recent developments in mathematical finance and financial econometrics and then summarizes the basic financial problems that motivate the SIML estimation in this book.

2.1 Developments in Quantitative Finance

In Figs. 2.1 and 2.2, we show daily data from the Nikkei-225 spot price index and the dollar–yen exchange rate during 2000~2016, which are two major sources of financial data in Japan. It is commonly observed that price of an actively traded (financial) commodity in a financial market fluctuates wildly over time, and predicting its future value based on present and past observations is known to be difficult. A basic and interesting question is how to interpret these price movements over time and among early attempts (Bachelier 1900) tried to develop a key mathematical idea to use Brownian motion as an appropriate mathematical tool.

Let the security price at $t \in [0, T]$ be $S(t)$, which is a continuous-time stochastic process. We divide $[0, T]$ into n intervals, $\Delta t = T/n$, and let $S_n(t_k^n)$ be the security price at $t_k^n \in ((k-1)T/n, kT/n]$, $k = 1, \ldots, n$. For simplicity, we often take $t_k^n = kT/n$, and we set $t_0^n = 0$ and $T = 1$. When there is no dividend in $[0, T]$, the returns of $S_n(t_i^n)$ can be defined by

$$r_n(i\,\Delta t) = \frac{S_n(t_k^n) - S_n(t_{k-1}^n)}{S_n(t_{k-1}^n)} \quad (i = 1, \ldots, n), \tag{2.1}$$

where $S_n(t_0^n)$ is the initial (fixed) price and $\Delta t = 1/n$.

For each return, we consider the simple situation when in the period $((i-1)\Delta t, i\Delta t]$, the return satisfies the next assumption.

© The Author(s) 2018
N. Kunitomo et al., *Separating Information Maximum Likelihood Method for High-Frequency Financial Data*, JSS Research Series in Statistics, https://doi.org/10.1007/978-4-431-55930-6_2

Fig. 2.1 Nikkei-225 spot price (daily closing data)

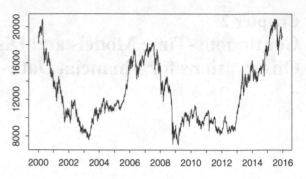

Fig. 2.2 Yen–dollar exchange rate (daily closing data)

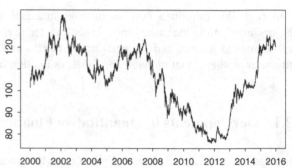

Assumption 2.1 For any $t \in [0, T]$ and $S_n(t) \ (> 0)$, $r_n(i\Delta t) \ (i = 1, \ldots, n)$ satisfy the following:

(i) the expected return is constant as $\mathbf{E}^P[r_n(i\Delta t)] = \mu \Delta t \ (i = 1, 2, \ldots, n)$;

(ii) $r_n(i\Delta t)$ and $r_n(j\Delta t) \ (i \neq j)$ are uncorrelated random variables;

(iii) the variances of returns are constant as $\mathbf{V}^P[r_n(i\Delta t)] = \sigma^2 \Delta t$;

(iv) the third-order moment exists and $\mathbf{E}^P[|r_n(i\Delta t)|^3] = o(\Delta t)$,

where $\mathbf{E}^P[\,\cdot\,]$ is the mathematical expectation under the probability measure P.

The parameters μ and σ are called the drift and volatility coefficients, respectively. The excess return $r_n^*(i\Delta t) = r(i\Delta t) - \mu \Delta t \ (i = 1, \ldots, n)$ is a sequence of martingale differences; that is, the conditional expectation $\mathbf{E}^P[r_n^*(i\Delta t)|\mathscr{F}_{n,i-1}] = 0 \ (i = 1, \ldots, n)$, where $\mathscr{F}_{n,i-1}$ is the information at $(i-1)\Delta t$ as the σ-field. Because we denote $r_n^*(i\Delta t) = r_n(i\Delta t) - \mu \Delta t$, then the conditional expectation $\mathbf{E}^P[r_n^*(i\Delta t)|r_n^{(*)}(j\Delta t), i - 1 \geq j] = 0$ and $\mathbf{E}^P[(r_n^*(i\Delta t))^2|r_n^*(j\Delta t), i - 1 \geq j] = \sigma^2 \Delta t$. From the sequence of mutually uncorrelated random variables, we define $\{B_n(i\Delta t)\}$ such that $r_n^*(i\Delta t) = \sigma[B_n(i\Delta t) - B_n((i-1)\Delta t)]$ and denote the initial value $B_n(0) = 0$. Then, we write $\sigma B_n(i\Delta t) = \sum_{j=1}^{i} r_n^*(j\Delta t)$ and $r_n(i\Delta t) = \mu \Delta t + \sigma[B_n(i\Delta t) - B_n((i-1)\Delta t)]$. We interpolate the price process $B_n(t)$ for any $t \in [0, T]$ by using the values at discrete points. Then,

$$\frac{S_n(i\Delta t)}{S_n(0)} = \frac{S_n(i\Delta t)}{S_n((i-1)\Delta t)}\frac{S_n((i-1)\Delta t)}{S_n((i-2)\Delta t)}\cdots\frac{S_n(\Delta t)}{S_n(0)}$$

$$= \prod_{j=1}^{i}(1 + r_n(j\Delta t)).$$

By taking the logarithm of both sides, we have for small Δt that $\log r_n(j\Delta t)) = r_n(j\Delta t) - \frac{1}{2}r_n(j\Delta t)^2 + o_p(\Delta t)$. By using the notation $[nt] = i(n)$,

$$\log[\frac{S_n(i(n)\Delta t)}{S(0)}] \sim \sum_{j=1}^{i(n)} r_n(j\Delta t) - \frac{1}{2}\sum_{j=1}^{i(n)} r_n(j\Delta t)^2$$

$$\sim \sum_{j=1}^{i(n)}\left\{\left(\mu - \frac{\sigma^2}{2}\right)\Delta t + \sigma[B_n(j\Delta t) - B_n((j-1)\Delta t)]\right\}$$

$$= \left(\mu - \frac{\sigma^2}{2}\right)[i(n)\Delta t] + \sigma B_n(i(n)\Delta t).$$

For any fixed $t \in [0, 1]$, as $n \to +\infty$, we have $\Delta t \to 0$. Hence, as $i(n)\Delta t \to t$, $S_n(t) \to S(t)$ and $B_n(t) \to B(t)$ in the sense of convergence in distribution. As the limit, we write

$$\log \frac{S(t)}{S(0)} = \left(\mu - \frac{\sigma^2}{2}\right)t + \sigma B(t). \tag{2.2}$$

Then, by using the functional central limit theorem (FCLT) (see Section 37 of Billingsley (1995) for instance), we summarize the above arguments.

Theorem 2.1 *Under Assumption 2.1, as $\Delta t \longrightarrow 0$ the security price process $\{S(t)\}$ satisfies the geometric Brownian motion (GMB) of (2.2) with the drift parameter μ and volatility parameter σ, where $\{B(t)\}$ is the standard Brownian motion (SBM) on $[0, T]$.*

The random variable $B(t)$ follows a normal distribution with zero mean and variance t. It is known that for any t, $B(t)$ does not have a finite variation and the sample path of Brownian motion is in $C[0, T]$, which is the totality of all continuous functions on $[0, T]$, but it is nowhere differentiable. The GBM satisfies the stochastic differential equation (SDE) given by

$$dS = \mu S dt + \sigma S dB. \tag{2.3}$$

2.2 On Financial Derivatives and the Black–Scholes formula

We consider the financial market in which (i) there is a risky security with price $S(t)$ at time t in the period $[0, T]$, (ii) there is also a safe asset without any risk (i.e., a government bond), and (iii) many participants can trade two assets at any time

$(t \in [0, T])$ with any arbitrary unit. As a friction-less financial market, we assume several conditions.

[**Condition 2.2**]: At any time $t \in [0, T]$, we can trade assets freely and continuously and the financial market satisfies the following conditions

(i) there is no transaction cost such as tax and trade commission, and short sales are possible,

(ii) the risky asset has no default risk, and the security price follows the GBM of (2.2). There is no dividend during the period, and we have a constant compound interest r (≥ 0) if we invest the safe asset in the period,

(iii) in the economy, there is no-arbitrage opportunity.

The economy satisfying the above conditions is called the Black–Scholes (BS) model (Black and Scholes 1972). When we have a safe asset with constant interest rate, we normalize the price of risky asset by the bond price e^{rt} (i.e., continuous compounding), and then the present value of the risky asset $X(t) = e^{-rt}S(t)$ satisfies

$$X(t) = X(0)e^{(\mu - r - \frac{\sigma^2}{2})t + \sigma B(t)}$$

and under the probability measure P. If we set $B^*(t) = B(t) + \left(\frac{\mu - r}{\sigma}\right)t$, then

$$X(t) = X(0)e^{(-\frac{\sigma^2}{2})t + \sigma B(t)^*} . \tag{2.4}$$

In this representation, we take another probability measure Q such that $\{B(t)^*\}$ is an SBM with Q. We displace the drift term under P by $(\mu - r)t/\sigma$. Let $\theta = (\mu - r)/\sigma$ and use the exponential martingale

$$M(t) = \exp\left[-\theta B(t) - \frac{1}{2}\theta^2 t \right]. \tag{2.5}$$

For any measurable set **A**, we define

$$Q(\mathbf{A}) = \mathbf{E}^P[M(T)1_{\{\mathbf{A}\}}]. \tag{2.6}$$

Then if we take $\mathbf{A} = \Omega$ (the sample space), we have $Q(\Omega) = 1$. Conversely, if we set any $\mathbf{A}$ $(\subset \Omega)$, then $Q(\mathbf{A}) > 0$ is a probability measure. When B^* is an SBM under Q, we have the martingale property under Q of $X(t)$ as

$$\mathbf{E}^Q[X(t)|X(r), t > s \geq r] = X(s) \ (a.s.). \tag{2.7}$$

Let a measurable set **A** be in $\mathscr{F}_s$ and a σ−field $\mathscr{F}_s$ can be interpreted as the information sets available for participants in a market at s. By using the conditional expectation operation, for any $0 \leq s \leq t$ and $\mathbf{A} \in \mathscr{F}_s$,

$$\begin{aligned}
\mathbf{E}^P[X(t)M(t)1_A] &= \int_A \mathbf{E}^P[X(t)M(t)|\mathscr{F}_s]1_A \, dP \\
&= \int_A X(s)dQ \\
&= \mathbf{E}^Q[X(t)|\mathscr{F}_s],
\end{aligned}$$

(2.8)

where we have used the product process $X(t)M(t)$ as

$$X(t)M(t) = X(0)\exp[(\sigma\theta - \frac{\sigma^2}{2})t + \sigma B(t)] \times \exp[-\theta B(t) - \frac{\theta^2}{2}t]$$

$$= X(0)\exp[(\sigma - \theta)B(t) - \frac{1}{2}(\sigma - \theta)^2 t].$$

Then, $\mathbf{E}^P[X(t)M(t)|\mathscr{F}_s] = X(s)M(s)$ $(t > s)$ and hence for $k = 1, 2$ we have

$$\begin{aligned}
\mathbf{E}^Q[B^*(t)^k] &= \mathbf{E}^Q[(B(t) + \theta t)^k] \\
&= \mathbf{E}^P[(B(t) + \theta t)^k e^{-\theta B(t) - \frac{1}{2}\theta^2 t}] \\
&= \int_{\mathbf{R}}(x + \theta t)^k e^{-\theta x - \frac{1}{2}\theta^2 t} \frac{1}{\sqrt{2\pi t}} e^{-\frac{1}{2t}x^2} dx.
\end{aligned}$$

For instance, we find that $\mathbf{E}^Q[B^*(t)] = 0$ and $\mathbf{E}^Q[(B^*(t))^2] = t$, and thus, B^* is a Q-SBM.

To summarize the basic arguments, the stochastic process $X(t)$ is a continuous martingale with respect to the probability measure Q, and by using this SBM under Q, we represent the risky asset price as the solution of stochastic differential equation (SDE)

$$dS = r \, S dt + \sigma \, S dB^*.$$

(2.9)

The important issue is that the drift term of the stochastic process μ has been replaced by the return of the riskless asset r, which corresponds to the no-arbitrage condition in the financial market.

As consequence, the theoretical price of the European-type option contract with pay-off function $g(S(T))$ at the expiration period T is given by the conditional expectation

$$V_t = \mathbf{E}_t^Q[e^{-r(T-t)}g(S(T))],$$

(2.10)

where $\mathbf{E}_t^Q[\cdot]$ is the conditional expectation given the information at time t. For instance, if we take an European call option contract with pay-off function $g(S(T)) = \max\{S(T) - K, 0\}$ (K is a positive constant), then it is possible to find a function (or strategy) $\pi^*(t)$ and v such that

$$g(S(T)) = v e^{r(T-t)} + \int_t^T e^{r(T-s)}\pi^*(s)\sigma \, dB^*(s).$$

(2.11)

As a typical exercise, we take the theoretical price of European call option when the price process follows GBM and there is no dividend in the period $(t, T]$. We denote the density of $S(t)$ at time t as $f(S(t))$, and then,

$$C(S, t) = \mathbf{E}^Q[e^{-r(T-t)} \max\{S(T) - K, 0\}|S(t) = S]$$
$$= e^{-r(T-t)} \int_K^\infty S(T)f(S(T))dS(T) - Ke^{-r(T-t)} \int_K^\infty f(S(T))dS(T).$$

We denote $\mu^* = r - \sigma^2/2$ and use the log transformation, whereupon the distribution at T is given by $X = \log S(T) \sim N[\log S + \mu^*(T - t), \sigma^2(T - t)]$. We set constants

$$d_1 = \frac{\log(S/K) + (r + \frac{\sigma^2}{2})(T - t)}{\sigma\sqrt{T - t}} \tag{2.12}$$

and $d_2 = d_1 - \sigma\sqrt{T - t}$. For the second integration, we can use the relation

$$\mathbf{P}(S(T) \geq K) = \mathbf{P}[\frac{\log S(T) - \log S - \mu^*(T - t)}{\sigma\sqrt{T - t}} \geq \frac{\log K - \log S - \mu^*(T - t)}{\sigma\sqrt{T - t}}]$$
$$= N(d_2),$$

where $N(\cdot)$ is the standard normal distribution function given by $N(z) = \int_{-\infty}^z \frac{1}{\sqrt{2\pi}} e^{-x^2/2}dx$, and we have $1 - N(-d_2) = N(d_2)$. The first integration becomes

$$\int_K^\infty S(T)f(S(T))dS(T)$$
$$= \int_{\log K}^\infty e^x \frac{1}{\sqrt{2\pi\sigma^2(T - t)}} \exp\{-\frac{[x - \log S - \mu^*(T - t)]^2}{2(T - t)\sigma^2}\}dx$$
$$= Se^{r(T-t)}N(d_1)$$

by using the factorization of the exponential part. Thus, we can summarize the fair-price as

$$C(S, t) = SN(d_1) - Ke^{-r(T-t)}N(d_2). \tag{2.13}$$

The above formula corresponds to the one derived by Black and Scholes (1973), who solved the partial differential equation (PDE)

$$\frac{1}{2}\sigma^2 S^2 C_{SS} + rSC_S + C_t - rC = 0, \tag{2.14}$$

subject to the boundary condition $C(S, T) = \max\{S(T) - K, 0\}$, where C_{SS} is a second-order partial derivative and C_S and C_t are the first-order partial derivatives.

2.3 Diffusion, Realized Volatility, and Micro-Market Noise

The GBM model has been successful as the classical model of the asset-price process in financial economics mainly because we have the explicit BS formula for derivative pricing. In this framework, the unknown drift parameter μ is replaced by r, which is observable, and the only unknown parameter we need is the volatility σ^2 (or σ) in (2.9), and we then need to estimate the volatility parameter from data.

However, it is quite a special case in the sense that both the drift and volatility parameters are constant over time as a continuous-time stochastic process. The obvious limitation of the GBM model in application is that the log-return follows the Gaussian distribution, but then there are many empirical studies against this aspect in real financial data; that is, we often find that the actual financial returns exhibit fat tails and non-Gaussianity.

As the theory of continuous-time stochastic processes, a more general form of the SDE is given by

$$dS = \mu_t S_t dt + \sigma_t S_t dB_t,\tag{2.15}$$

which has been called the diffusion-type continuous process. (In stochastic analysis, the notation with $\mu_t^* = \mu_t S_t$ and $\sigma_t^* = \sigma_t S_t$ is often used. There is no essential difference due to Ito's Lemma.) The GBM model corresponds to a special case in which $\mu_t = \mu$ and $\sigma_t = \sigma$ with constants μ and σ. (2.15) has another representation as

$$S(t) = S(0) + \int_0^t \mu_s S_s ds + \int_0^t \sigma_s S_s dB_s,\tag{2.16}$$

where the first term is an integration in the sense of Riemann, while the second term is an Itô's stochastic integration with respect to the Brownian motion B_t. A detailed theory of stochastic differential equation (SDE) and stochastic integration has been explained by Ikeda and Watanabe (1989).

In order to estimate the volatility parameter, it may be natural to use the set of returns (2.1) from a set of n observations of the underlying prices. When we have high-frequency data, n can be large. From the traditional statistical theory, we can estimate the unknown parameters precisely as we have finer observations. For the simplicity of our arguments, we set $p = 1$ (one-dimensional case), $T = 1$, and $\Delta t = 1/n$. For the observed prices $S(t_i^n)$, which follows the continuous-time diffusion process, we define the log-returns $Y(t_i^n) = \log S(t_i^n)$ (or $S(t_i^n)$). Then, the realized volatility (RV) can be defined by

$$RV_n = \sum_{i=1}^n [Y(t_i^n) - Y(t_{i-1}^n)]^2\tag{2.17}$$

Fig. 2.3 x (time in second)
versus y (realized volatility)

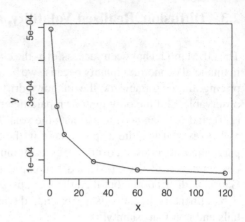

(or we use $\sum_{i=1}^{n} [Y(t_i^n)/Y(t_{i-1}^n) - 1]^2$ for $Y(t_i^n) = S(t_i^n)$ in (2.17)), and it is known in stochastic analysis

$$\text{RV}_n \xrightarrow{p} \int_0^1 \sigma_s^2 ds \qquad (2.18)$$

as $n \longrightarrow \infty$ in the sense of convergence in probability for the class of continuous-time diffusion processes. (See Chap. 5, for instance.)

When $\sigma_s = \sigma$ (a constant parameter), the probability limit of (2.18) is σ^2, which was originally called the volatility or diffusion parameter because of Black–Scholes derivatives theory. Then, there were a number of applications using RV, which measures the risk involved in many financial markets. We mainly use the log-returns ($Y(t_i^n) = \log S(t_i^n)$) for the resulting convenience in the following analysis.

However, it is known that there is a strong evidence against the underlying argument in this line of reasoning. For instance, when we take the estimates of the realized volatilities with different n by using high-frequency data, we find that the estimate of σ^2 becomes large as we take finer high-frequency data (from 120 to 1 s). As an example, we have Fig. 2.3, which gives the realized volatility of Nikkei-225 Futures based on Table 4.1. (We shall discuss this application more in Chap. 4.) When we take high-frequency financial data, the effects of micro-market noise cannot be negligible. In Fig. 2.4, we give the high-frequency data (from 1 to 60 s) of Nikkei-225 Futures in one day of the year 2007. These figures also suggest that the observed sample paths are not realizations of continuous-time diffusion processes, which were developed in stochastic analysis.

As a possible explanation, we consider the role of noise in high-frequency financial data. For this purpose, first, we consider the additive model for the observed (log-)price at $t_i^n \in [0, 1]$ as

$$Y(t_i^n) = X(t_i^n) + \varepsilon_n v(t_i^n) \quad (i = 1, \ldots, n), \qquad (2.19)$$

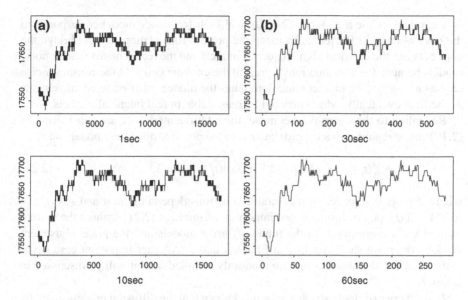

Fig. 2.4 a : Nikkei-225 Futures (1 and 10 min). **b** : Nikkei-225 Futures (30 and 60 min)

where ε_n is a sequence of nonnegative numbers and $X(t)$ is the continuous-time Brownian martingale as the simplest case with

$$X(t) = X(0) + \int_0^t \sigma_s dB_s \ (0 \leq s \leq 1).$$ (2.20)

We take B_s as SBM, and σ_s is the (instantaneous) volatility function, which is predictable (and progressively measurable) with respect to $(\mathbf{\Omega}, \mathscr{F}, (\mathscr{F}_t)_{t \geq 0}, P)$. We use the notations $X_i = X(t_i^n)$ and $Y_i = Y(t_i^n)$, where $Y_i = \log S(t_i^n)$ in the following one-dimensional analysis.

For instance, ε_n can be a constant, and then, the market noise term dominates the realized volatility as $n \to \infty$. When ε_n are small, however, there can be other possibilities.

In this way, we can construct the diffusion model with micro-market noise, which gives an important way to resolve the contradictory evidence on the estimation of volatility. Then, it may be important to investigate the situation in which the micro-market noise terms $v(t_i^n) (= v_i)$ as a sequence of random variables with $\mathbf{E}[v_i] = 0, \mathbf{E}[v_i^2] = 1$, and $\varepsilon_n \ (\geq 0)$ is a (nonnegative) sequence of parameters depending on n, which goes to zero as $n \longrightarrow \infty$. We call this situation the small-noise case. For instance, Kunitomo and Kurisu (2017), and Kurisu (2017) have used (2.19) when ε_n is a sequence of constants depending on n, which goes to zero as $n \to \infty$. They showed that the standard methods for estimating volatility parameter, such as the realized volatility, are quite sensitive to the presence of micro-market noise. The standard model of micro-market noise in many studies is the case when $\varepsilon_n = \varepsilon \ (>$

0, a constant) while $n \longrightarrow \infty$. The classical high-frequency model corresponds to the case when $\varepsilon_n = 0$. Hence, by using the present formulation, we can bridge the gaps between the classical high-frequency models and the recent micro-market noise models because the two situations represent the extreme cases. In the constant-noise case as $n \longrightarrow \infty$, the market noise dominates the hidden intrinsic price movements as the limit eventually, which may not be reasonable in real financial markets.

Regarding to the micro-market noise, the additive model we are considering in (2.19) can be regarded as an approximation to the possible nonlinear models, such as

$$Y(t_i^n) = f_n(X(t_i^n), \Delta X(t_i^n), v(t_i^n)) \quad (i = 1, \dots, n), \tag{2.21}$$

where $f_n(\cdot)$ is a sequence of measurable functions depending on n and $\Delta X(t_i^n) = X(t_i^n) - X(t_{i-1}^n)$. For instance, one important example of (2.21) without the second term $\Delta X(t_i^n)$ corresponds to the round-off error models and the price-adjustment models which we shall consider in Chaps. 6 and 7. Another important case is the situation when the observations are randomly sampled, and it will be discussed in Chap. 7.

In the theory of stochastic analysis, it is known that the diffusion process given by (2.15) and (2.16) can be extended to the class of Itô semi-martingales. When jumps are possible in the underlying stochastic process and $p = 1$, it is natural to assume that the hidden price process follows an Itô's semi-martingale (continuous-time) process

$$X(t) = X(0) + \int_0^t \mu_s ds + \int_0^t \sigma_s dB_s + \int_0^t \int_{|\delta(s,x)|<1} \delta(s, x)(\mu - v)(ds, dx) \\ + \int_0^t \int_{|\delta s,x)|\geq 1} \delta(s, x)\mu(ds, dx) , \tag{2.22}$$

where μ_s (drift parameter) and σ_s (diffusion parameter) are bounded, predictable, and progressively measurable, $\delta(s, x)$ is a predictable process, $\mu(\cdot)$ is a jump measure, and $v(\cdot)$ is the compensator of $1_A * \mu$ for $1 * v(\omega)_t = v(\omega : [0, t) \times A)$. Here, we follow the notation of Ikeda and Watanabe (1989, Chap. 2), Jacod and Protter (2012, Section 2), and the details of diffusion and jump processes are explained rigorously in these standard textbooks on stochastic analysis.

Here, we need to formulate the problem of micro-market noise in financial markets with some mathematical notation. Let the first-filtered probability space be $(\Omega^{(0)}, \mathcal{F}^{(0)}, (\mathcal{F}_t^{(0)})_{t\geq 0}, P^{(0)})$ on which the Itô semi-martingale X_t $(0 \leq t \leq 1)$ is well-defined. Let also the second-filtered probability space be $(\Omega^{(1)}, \mathcal{F}^{(1)}, (\mathcal{F}_t^{(1)})_{t\geq 0}, P^{(1)})$ on which the micro-market noise terms $v(t_i^n)$ $(i = 1, \dots, n)$ are well-defined with $0 \leq t_i^n \leq 1$. Then, we can construct the filtered probability space and the probability measure as $(\Omega, \mathcal{F}, (\mathcal{F}_t)_{t\geq 0}, P)$, where $\Omega = \Omega^{(0)} \times \Omega^{(1)}$, $\mathcal{F} = \mathcal{F}^{(0)} \otimes \mathcal{F}^{(1)}$ with $\mathcal{F}_t = \bigcap_{s>t} \mathcal{F}_s^{(0)} \otimes \mathcal{F}_s^{(1)}$ $(0 \leq t \leq s \leq 1)$ and $P = P^{(0)} \times P^{(1)}$.

In the next chapter, we extend the discussions of the one-dimensional diffusion case in this chapter to general p-dimensional cases with micro-market noise. For ease of exposition, we make several assumptions about the underlying stochastic processes, but most of them can be extended to the more general cases. For instance, we treat a further problem of jumps in Chap. 9. However, we will not purse mathematical rigor because of their complicated arguments involved. We generally require few conditions essentially at the end, but we omit the detailed mathematical developments.

References

Bachelier, L. 1900. Théorie de Spéculation. In *A translation in random character of stock prices*, ed. P. Coortner, 17–79. MIT Press.
Billingsley, P. 1995. *Probability and measure*, 3rd ed. New York: Wiley.
Black, F., and M. Scholes. 1972. The pricing options and corporate liabilities. *Journal of Political Economy* 81: 7637–659.
Ikeda, N., and S. Watanabe. 1989. *Stochastic differential equations and diffusion processes*, 2nd ed. North-Holland.
Kunitomo, N., and D. Kurisu. 2017. Effects of jumps and small noise in high-frequency financial econometrics. *Asia-Pasific Financial Markets* 24: 39–73.
Kurisu, D. 2017. Power variations and testing for co-jumps: the small noise approach. *Scandinavian Journal of Statistics*. http://onlinelibrary.wiley.com/doi/10.1111/sjos.12309/abstract.
Jacod, J., and P. Protter. 2012. *Discretization of Processes*. Berlin: Springer.

Chapter 3
The SIML Estimation of Volatility and Covariance with Micro-market Noise

Abstract We introduce the SIML method for estimating the integrated volatility and co-volatility (or covariance) parameters from a set of discrete observations. We first define the SIML estimator in the basic case and then give the asymptotic properties of the SIML estimator in more general cases.

3.1 Statistical Models in Continuous-Time and Discrete-Time

Let y_{ij} be the ith observation of the jth (log-) price at t_i^n for $i = 1, \ldots, n$; $j = 1, \ldots, p$; $0 = t_0^n \leq t_1^n \leq \cdots \leq t_n^n = 1$. We set $\mathbf{y}_i = (y_{i1}, \ldots, y_{ip})'$ be a $p \times 1$ vector, and $\mathbf{Y}_n = (\mathbf{y}_i')$ is an $n \times p$ matrix of observations and $\mathbf{y}_0$ is the initial observation vector. The underlying continuous vector process $\mathbf{x}_i$ at t_i^n $(i = 1, \ldots, n)$ is not necessarily the same as the observed prices, and let $\mathbf{v}_i' = (v_{i1}, \ldots, v_{ip})$ be the vector of the additive micro-market noise at t_i^n, which is independent of $\mathbf{x}_i$. Then we have

$$\mathbf{y}_i = \mathbf{x}_i + \mathbf{v}_i , \qquad (3.1)$$

where $\mathbf{v}_i$ forms a sequence of independent random variables with $E(\mathbf{v}_i) = \mathbf{0}$ and $E(\mathbf{v}_i \mathbf{v}_i') = \boldsymbol{\Sigma}_v$. We assume that $\boldsymbol{\Sigma}_v$ is nonnegative definite and finite. We focus on the equidistance case, and we set $h_n = t_i^n - t_{i-1}^n = 1/n$ $(i = 1, \ldots, n)$ in this chapter. We assume that

$$\mathbf{x}(t) = \mathbf{x}_0 + \int_0^t \mathbf{C}_x(s) d\mathbf{B}(s) \quad (0 \leq t \leq 1), \qquad (3.2)$$

where we denote $\mathbf{x}_i = \mathbf{x}(t_i^n)$ $(i = 1, \ldots, n)$ for convenience, $\mathbf{B}(s)$ is a $q \times 1$ $(q \geq 1)$ vector of the standard Brownian motions, and $\mathbf{C}_x(s) = (c_{gh}^{(x)}(s))$ is a $p \times q$ (bounded) coefficient matrix that is progressively measurable with respect to $\mathscr{F}_s$ $(s \geq 0)$ and predictable, and $\mathscr{F}_0$ is the initial σ-field. We write the instantaneous diffusion function $\boldsymbol{\Sigma}_x(s)$ $(= (\sigma_{gh}^{(x)}(s))) = \mathbf{C}_x(s)\mathbf{C}_x'(s)$, where $\mathbf{C}_x'(s)$ is the transposed matrix of $\mathbf{C}_x(s)$ and $\mathscr{F}_s$ is the σ-field generated by $\{\mathbf{B}(r), r \leq s\}$.

© The Author(s) 2018

N. Kunitomo et al., *Separating Information Maximum Likelihood Method for High-Frequency Financial Data*, JSS Research Series in Statistics, https://doi.org/10.1007/978-4-431-55930-6_3

The main statistical problem is to estimate the integrated volatility and co-volatility as

$$\Sigma_x = (\sigma_{gh}^{(x)}) = \int_0^1 \Sigma_x(s)ds \tag{3.3}$$

of the underlying continuous process $\{\mathbf{x}(t)\}$ and the variance–covariance $\Sigma_v = (\sigma_{gh}^{(v)})$ of the noise process from the observed discrete-time process $\mathbf{y}_i$ ($i = 1, \ldots, n$). We use the notation that $\sigma_{gh}^{(x)}(s)$ and $\sigma_{gh}^{(v)}$ are the (g, h)-th element of $\Sigma_x(s)$ and Σ_v, respectively. When $p = q = 1$, we sometimes use the notation $\mathbf{C}_x(s) = \sigma_x(s)$ (or σ_s) and $\Sigma_x = \sigma_x^2$ for convenience.

There are three different situations regarding the instantaneous variance–covariance function, and we discuss the estimation problem of each. (i) When the coefficient matrix is constant, i.e., $\mathbf{C}_x(s) = \mathbf{C}$, we call this the basic case or simple case. The instantaneous variance and covariance are constant over time, and then the integrated variance and covariance are constant. (ii) When the coefficient matrix varies with time, but it is a deterministic function of time ($\mathbf{C}_x(s)$), we call this the deterministic time-varying case. (iii) When the coefficient matrix varies with time and it is a stochastic function of time ($\mathbf{C}_x(s)$), we call this the stochastic case. For case (iii), we denote the conditional covariance function of the (underlying) price returns without micro-market noise as

$$\mathbf{E}\left[(\mathbf{x}_i - \mathbf{x}_{i-1})(\mathbf{x}_i - \mathbf{x}_{i-1})' | \mathscr{F}_{n,i-1}\right] = \int_{t_{i-1}}^{t_i} \mathbf{E}_{i-1}[\Sigma_x(s)]ds \ ,$$

where $\mathbf{r}_i = \mathbf{x}_i - \mathbf{x}_{i-1}$ is a sequence of martingale differences, $\mathbf{E}_{i-1}[\Sigma_x(s)]$ is the time-dependent (instantaneous) conditional variance–covariance matrix, $\mathscr{F}_{n,i-1}$ is the σ-field generated by $\mathbf{x}_j$ ($j \leq i - 1$) with (3.1) and $\mathbf{v}_j$ ($j \leq i - 1$), and $\mathscr{F}_{n,0}$ is the initial σ-field. Without loss of generality, we assume that $\Sigma_x(s)$ is a progressively measurable instantaneous variance–covariance matrix and $\sup_{0 \leq s \leq 1} \| \Sigma_x(s) \| < \infty$ (a.s.).

3.2 Basic Case

We first consider the simple situation in which $\mathbf{x}_i, \mathbf{v}_i$ ($i = 1, \ldots, n$) are independent with $\Sigma_x(s) = \Sigma_x$ ($0 \leq s \leq 1$), and $\mathbf{v}_i$ are independently, identically, and normally distributed as $N_p(\mathbf{0}, \Sigma_v)$. Then $\Delta \mathbf{x}_i = \mathbf{x}_i - \mathbf{x}_{i-1}$ follows $N_p(\mathbf{0}, (1/n)\Sigma_x)$. We use an $n \times p$ matrix $\mathbf{Y} = (\mathbf{y}_i')$ and consider the distribution of the $np \times 1$ random vector $(\mathbf{y}_1', \ldots, \mathbf{y}_n')'$. Given the initial condition $\mathbf{y}_0$, we have

$$\mathbf{Y}_n \sim N_{n \times p}\left(\mathbf{1}_n \cdot \mathbf{y}_0', \mathbf{I}_n \otimes \Sigma_v + \mathbf{C}_n \mathbf{C}_n' \otimes h_n \Sigma_x\right) , \tag{3.4}$$

where $\mathbf{1}_n' = (1, \ldots, 1)$, $\mathbf{I}_n$ is the $n \times n$ identity matrix, $h_n = 1/n$ ($= t_i^n - t_{i-1}^n$), $\mathbf{C}_n'$ is the transposed matrix of $\mathbf{C}_n$ and

$$\mathbf{C}_n = \begin{pmatrix} 1 & 0 & \ldots & 0 & 0 \\ 1 & 1 & 0 & \ldots & 0 \\ 1 & 1 & 1 & \ldots & 0 \\ 1 & \ldots & 1 & 1 & 0 \\ 1 & \ldots & 1 & 1 & 1 \end{pmatrix} . \tag{3.5}$$

To investigate the likelihood function in the basic case, we prepare the next lemma, which may be of independent interest. The proof is given in Chap. 5.

Lemma 3.1 *(i) Define an $n \times n$ matrix $\mathbf{A}_n$ by*

$$\mathbf{A}_n = \frac{1}{2} \begin{pmatrix} 1 & 1 & 0 & \ldots & 0 \\ 1 & 0 & 1 & \ldots & 0 \\ 0 & 1 & 0 & 1 & \ldots \\ 0 & 0 & \ldots & 0 & 1 \\ 0 & \ldots & 0 & 1 & 0 \end{pmatrix} . \tag{3.6}$$

Then $\cos \pi (\frac{2k-1}{2n+1})$ $(k = 1, \ldots, n)$ are eigenvalues of $\mathbf{A}_n$, and the eigenvectors are

$$\begin{bmatrix} \cos \left[\pi \left(\frac{2k-1}{2n+1} \right) \frac{1}{2} \right] \\ \cos \left[\pi \left(\frac{2k-1}{2n+1} \right) \frac{3}{2} \right] \\ \vdots \\ \cos \left[\pi \left(\frac{2k-1}{2n+1} \right) \left(n - \frac{1}{2} \right) \right] \end{bmatrix} \quad (k = 1, \ldots, n). \tag{3.7}$$

(ii) We have the spectral decomposition

$$\mathbf{C}_n^{-1} \mathbf{C}_n'^{-1} = \mathbf{P}_n \mathbf{D}_n \mathbf{P}_n' = 2\mathbf{I}_n - 2\mathbf{A}_n , \tag{3.8}$$

where $\mathbf{D}_n$ is a diagonal matrix with the kth element

$$d_k = 2 \left[1 - \cos \left(\pi \left(\frac{2k-1}{2n+1} \right) \right) \right] \quad (k = 1, \ldots, n) , \tag{3.9}$$

$$\mathbf{C}_n^{-1} = \begin{pmatrix} 1 & 0 & \ldots & 0 & 0 \\ -1 & 1 & 0 & \ldots & 0 \\ 0 & -1 & 1 & 0 & \ldots \\ 0 & 0 & -1 & 1 & 0 \\ 0 & 0 & 0 & -1 & 1 \end{pmatrix} \tag{3.10}$$

and

$$\mathbf{P}_n = (p_{jk}) , \quad p_{jk} = \sqrt{\frac{2}{n + \frac{1}{2}}} \cos \left[\frac{2\pi}{2n+1} \left(k - \frac{1}{2} \right) \left(j - \frac{1}{2} \right) \right] . \tag{3.11}$$

We transform $\mathbf{Y}_n$ to $\mathbf{Z}_n$ $(= (\mathbf{z}_k'))$ by

$$\mathbf{Z}_n = h_n^{-1/2} \mathbf{P}_n \mathbf{C}_n^{-1} \left(\mathbf{Y}_n - \bar{\mathbf{Y}}_0 \right) \tag{3.12}$$

where

$$\bar{\mathbf{Y}}_0 = \mathbf{1}_n \cdot \mathbf{y}_0' . \tag{3.13}$$

Given the initial condition $\mathbf{y}_0$, this transformation is one-to-one and all components of $\mathbf{Z}_n$ are independent in the present situation. By using the relation $d_k = 4 \sin^2(\pi/2)[(2k-1)/(2n+1)]$ $(k = 1, \ldots, n)$, the (conditional) likelihood function under Gaussian noise is given by

$$L_n^*(\boldsymbol{\theta}) = \left(\frac{1}{\sqrt{2\pi}} \right)^{np} \prod_{k=1}^{n} |a_{kn} \boldsymbol{\Sigma}_v + \boldsymbol{\Sigma}_x|^{-1/2} e^{\left\{ -\frac{1}{2} \mathbf{z}_k' \left(a_{kn} \boldsymbol{\Sigma}_v + \boldsymbol{\Sigma}_x \right)^{-1} \mathbf{z}_k \right\}} , \tag{3.14}$$

where we denote

$$a_{kn} = 4n \sin^2 \left[\frac{\pi}{2} \left(\frac{2k-1}{2n+1} \right) \right] \ (k = 1, \ldots, n) . \tag{3.15}$$

Hence, the (conditional) ML estimator can be defined as the solution of maximizing

$$L_n(\boldsymbol{\theta}) = \sum_{k=1}^{n} \log |a_{kn} \boldsymbol{\Sigma}_v + \boldsymbol{\Sigma}_x|^{-1/2} - \frac{1}{2} \sum_{k=1}^{n} \mathbf{z}_k' [a_{kn} \boldsymbol{\Sigma}_v + \boldsymbol{\Sigma}_x]^{-1} \mathbf{z}_k . \tag{3.16}$$

From this representation, we find that the ML estimator of unknown parameters is a rather complicated function of all observations in general because each a_{kn} term depends on k as well as n.

Let denote $a_{k_n,n}$ and k depending on n explicitly, whereupon we find that $a_{k_n,n} \to 0$ as $n \to \infty$ when $k_n = O(n^\alpha)$ $(0 < \alpha < \frac{1}{2})$ because $\sin x \sim x$ as $x \to 0$. Also $a_{n+1-l_n,n} = O(n)$ when $l_n = O(n^\beta)$ $(0 < \beta < 1)$. When k_n is small, we expect $a_{k_n,n}$ to be small. We denote a_{kn} without any confusion in the following discussion.

Then we may approximate $2 \times L_n(\boldsymbol{\theta})$ by

$$L_n^{(1)}(\boldsymbol{\theta}) = -m \log |\boldsymbol{\Sigma}_x| - \sum_{k=1}^{m} \mathbf{z}_k' \boldsymbol{\Sigma}_x^{-1} \mathbf{z}_k . \tag{3.17}$$

It is proportional to the standard log-likelihood function except for the fact that we use only the first m terms (see Lemma 3.2.2 of Anderson 2003). Then the SIML estimator of $\boldsymbol{\Sigma}_x$ is defined by

$$\hat{\boldsymbol{\Sigma}}_x = \frac{1}{m_n} \sum_{k=1}^{m_n} \mathbf{z}_k \mathbf{z}_k' . \tag{3.18}$$

By contrast, when l_n is small and $k_n = n + 1 - l_n$, we expect $a_{n+1-l_n,n}$ to be large. Thus we may approximate $2 \times L_n(\boldsymbol{\theta})$ by

$$L_n^{(2)}(\boldsymbol{\theta}) = - \sum_{k=n+1-l}^{n} \log |a_{kn} \boldsymbol{\Sigma}_v| - \sum_{k=n+1-l}^{n} \mathbf{z}_k'[a_{kn} \boldsymbol{\Sigma}_v]^{-1} \mathbf{z}_k . \tag{3.19}$$

This is also proportional to the standard log-likelihood function approach except for the fact that we use only the last l terms of (3.16). Then the SIML estimator of $\boldsymbol{\Sigma}_v$ is defined by

$$\hat{\boldsymbol{\Sigma}}_v = \frac{1}{l_n} \sum_{k=n+1-l_n}^{n} a_{kn}^{-1} \mathbf{z}_k \mathbf{z}_k' . \tag{3.20}$$

When $p = q = 1$, we use the notation σ_x^2 and $\hat{\sigma}_x^2$ for $\boldsymbol{\Sigma}_x$ and $\hat{\boldsymbol{\Sigma}}_x$, respectively, and also σ_v^2 and $\hat{\sigma}_v^2$ for $\boldsymbol{\Sigma}_v$ and $\hat{\boldsymbol{\Sigma}}_v$, respectively.

For both $\hat{\boldsymbol{\Sigma}}_v$ and $\hat{\boldsymbol{\Sigma}}_x$, the respective number of terms m_n and l_n can depend on n. Then we need only the order requirements that $m_n = O(n^\alpha)$ $(0 < \alpha < \frac{1}{2})$ and $l_n = O(n^\beta)$ $(0 < \beta < 1)$ for $\boldsymbol{\Sigma}_x$ and $\boldsymbol{\Sigma}_v$, respectively. In the above construction, we define the SIML estimator by approximating the exact likelihood function under Gaussian micro-market noise. Although the convergence rate of the estimator of the integrated volatility and covariance is not optimal when the volatility is constant (see Gloter and Jacod 2001), the SIML estimator has some asymptotic robustness as we will show in later chapters. The most important characteristic of the SIML estimator is its simplicity, and it has other properties for dealing with high-frequency data. The simplicity of the SIML method differs from other estimation methods known, and it is crucial because the number of observations of tick data becomes large in the standard statistical sense. It is quite easy to deal with the multivariate high-frequency data in our approach.

By using a linear transformation in (3.12) (and Lemma 5.2 in Chap. 5), we can alternatively write

$$\hat{\boldsymbol{\Sigma}}_x = \frac{1}{m} \left(\frac{2n}{n + \frac{1}{2}} \right) \sum_{k=1}^{m} \left[\sum_{i=1}^{n} \mathbf{r}_i^* \cos \left[\pi \left(\frac{2k-1}{2n+1} \right) \left(i - \frac{1}{2} \right) \right] \right]$$

$$\times \left[\sum_{j=1}^{n} \mathbf{r}_j^{*'} \cos \left[\pi \left(\frac{2k-1}{2n+1} \right) \left(j - \frac{1}{2} \right) \right] \right]'$$

$$= \sum_{i=1}^{n} c_{ii}^* \mathbf{r}_i^* \mathbf{r}_i^{*'} + \sum_{i \neq j} c_{ij}^* \mathbf{r}_i^* \mathbf{r}_j^{*'} ,$$

where $\mathbf{r}_i^* = \mathbf{y}_i - \mathbf{y}_{i-1}$ and

$$c_{ii}^* = \left(\frac{2n}{2n+1}\right)\left[1 + \frac{1}{m}\frac{\sin 2\pi m\left(\frac{i-1/2}{2n+1}\right)}{\sin\left(\pi\frac{i-1/2}{2n+1}\right)}\right],$$

$$c_{ij}^* = \frac{1}{2m}\left(\frac{2n}{2n+1}\right)\left[\frac{\sin 2\pi m\left(\frac{i+j-1}{2n+1}\right)}{\sin\left(\pi\frac{i+j-1}{2n+1}\right)} + \frac{\sin 2\pi m\left(\frac{j-i}{2n+1}\right)}{\sin\left(\pi\frac{j-i}{2n+1}\right)}\right] \quad (i \neq j).$$

Hence, we have an alternative representation of the SIML estimator in terms of returns (i.e., $\mathbf{y}_i - \mathbf{y}_{i-1} = (y_{i,j} - y_{i-1,j})$ with the observation interval h_n $(= 1/n)$).

3.3   Asymptotic Properties of the SIML Estimator in the Basic Case

Because the SIML estimator has a simple representation, it is not difficult to derive its asymptotic properties. To make our arguments clear, we first consider the asymptotic normality of the SIML estimator of integrated volatility and integrated covariance in the basic case and then the time-varying cases. It may be appropriate here to stress the fact that we do not assume that the noise process is Gaussian when developing the analysis of the asymptotic properties of the SIML estimator in this subsection.

Let $\mathbf{r}_i = \mathbf{x}_i - \mathbf{x}_{i-1}$ $(i = 1, \ldots, n)$. When $\mathbf{C}_x(s)$ $(0 \leq s \leq 1)$ does not depend on s, we write $\mathbf{C}_x$ $(= \mathbf{C}_x(s))$. The conditional covariance matrix is then given by

$$\mathbf{E}\left[n\,\mathbf{r}_i\mathbf{r}_i'|\mathscr{F}_{n,i-1}\right] = \mathbf{\Sigma}_x \tag{3.21}$$

for all i $(i = 1, \ldots, n)$, and $\mathscr{F}_{n,i-1}$ is the σ-field available at t_{i-1}^n. $(\mathscr{F}_{n,0}$ is the null-set.) The covariance matrix $\mathbf{\Sigma}_x = (\sigma_{gh}^{(x)})$ is a constant (nonnegative definite) matrix. Then we have the next result, the proof of which is given in Chap. 5.

Theorem 3.1 *We assume that $\mathbf{x}_i$ and $\mathbf{v}_i$ $(i = 1, \ldots, n)$ are mutually independent and follow (3.1) and (3.2) with $\mathbf{C}_x(s) = \mathbf{C}_x$, $\mathbf{\Sigma}_x(s) = \mathbf{C}_x\mathbf{C}_x' = \mathbf{\Sigma}_x \geq 0$ (nonnegative definite) for $s \in [0, 1]$ and $\mathbf{\Sigma}_v \geq 0$. We further assume that $\mathbf{v}_i$ are a sequence of independent random vectors with $\mathscr{E}[v_{ig}^2 v_{jh}^2] < \infty$ $(i, j = 1, \ldots, n; g, h = 1, \ldots, p)$. Define the SIML estimator $\hat{\mathbf{\Sigma}}_x = (\hat{\sigma}_{gh}^{(x)})$ of $\mathbf{\Sigma}_x = (\sigma_{gh}^{(x)})$ and $\hat{\mathbf{\Sigma}}_v = (\hat{\sigma}_{gh}^{(v)})$ of $\mathbf{\Sigma}_v = (\sigma_{gh}^{(v)})$ by (3.18) and (3.20), respectively.*

(i) For $m_n = [n^\alpha]$ and $0 < \alpha < 1/2$, as $n \longrightarrow \infty$

$$\hat{\mathbf{\Sigma}}_x - \mathbf{\Sigma}_x \xrightarrow{p} \mathbf{0}. \tag{3.22}$$

(ii) For $m_n = [n^\alpha]$ and $0 < \alpha < 0.4$, as $n \longrightarrow \infty$

$$\sqrt{m_n}\left[\hat{\sigma}_{gh}^{(x)} - \sigma_{gh}^{(x)}\right] \xrightarrow{w} N\left(0, \sigma_{gg}^{(x)}\sigma_{hh}^{(x)} + \left[\sigma_{gh}^{(x)}\right]^2\right). \qquad (3.23)$$

The covariance of the limiting distributions of $\sqrt{m_n}[\hat{\sigma}_{gh}^{(x)} - \sigma_{gh}^{(x)}]$ and $\sqrt{m_n}[\hat{\sigma}_{kl}^{(x)} - \sigma_{kl}^{(x)}]$ is given by $\sigma_{gk}^{(x)}\sigma_{hl}^{(x)} + \sigma_{gl}^{(x)}\sigma_{hk}^{(x)}$ ($g, h, k, l = 1, \ldots, p$).
(iii) For $l_n = [n^\beta]$ and $0 < \beta < 1$, as $n \longrightarrow \infty$

$$\hat{\Sigma}_v - \Sigma_v \xrightarrow{P} O. \qquad (3.24)$$

(iv) Furthermore

$$\sqrt{l_n}\left[\hat{\sigma}_{gh}^{(v)} - \sigma_{gh}^{(v)}\right] \xrightarrow{w} N\left(0, \sigma_{gg}^{(v)}\sigma_{hh}^{(v)} + \left[\sigma_{gh}^{(v)}\right]^2\right). \qquad (3.25)$$

The covariance of the limiting distributions of $\sqrt{l_n}[\hat{\sigma}_{gh}^{(v)} - \sigma_{gh}^{(v)}]$ and $\sqrt{l_n}[\hat{\sigma}_{kl}^{(v)} - \sigma_{kl}^{(v)}]$ is given by $\sigma_{gk}^{(v)}\sigma_{hl}^{(v)} + \sigma_{gl}^{(v)}\sigma_{hk}^{(v)}$ ($g, h, k, l = 1, \ldots, p$).

It may be obvious that we have the joint normality of $\hat{\Sigma}_x$ and $\hat{\Sigma}_v$ as the limiting distributions of the SIML estimator in this case if we take a closer look at the proofs in Chap. 5. One interesting observation is that the asymptotic covariance in (3.23) and (3.25) does not depend on the fourth-order moments of the non-normal noise. This feature has an important implication for the statistical inference such as the problems of constructing confidence interval and conducting hypothesis testing on the integrated volatility in the presence of micro-market noise. In the SIML approach, the testing procedures and confidence regions can be constructed rather directly by using (3.25) for the covariance of the underlying continuous-time stochastic process and the covariance of the noise.

In the decomposition

$$\frac{1}{n}\sum_{k=1}^{n}\mathbf{z}_k\mathbf{z}_k' = \left(\frac{m}{n}\right)\frac{1}{m}\sum_{k=1}^{m}\mathbf{z}_k\mathbf{z}_k' + \left(\frac{n-l-m}{n}\right)\frac{1}{n-l-m}\sum_{k=m+1}^{n-l}\mathbf{z}_k\mathbf{z}_k'$$
$$+ \left(\frac{l}{n}\right)\frac{1}{l}\sum_{k=n+1-l}^{n}\mathbf{z}_k\mathbf{z}_k',$$

three terms are asymptotically independent, and we can construct the testing procedure and confidence region on any elements of Σ_x and Σ_v based on them. One simple statistical testing example is to test the null-hypothesis $H_0 : \sigma_{gg}^{(v)} = 0$ versus $H_1 : \sigma_{gg}^{(v)} > 0$ for some g, where $\sigma_{gg}^{(v)}$ is the (g, g)-th element of Σ_v ($g = 1, \ldots, p$). For this problem, we consider the test statistic

$$T_1 = \sqrt{m_n} \left[\frac{\dfrac{1}{l_n} \displaystyle\sum_{k=n+1-l_n}^{n} z_{kg}^2}{\dfrac{1}{m_n} \displaystyle\sum_{k=1}^{m_n} z_{kg}^2} - 1 \right] , \tag{3.26}$$

where $\mathbf{z}_k = (z_{kg})$ $(k = 1, \ldots, n; g = 1, \ldots, p)$.

When H_0 is true, both $(1/l_n) \sum_{k=n+1-l_n}^{n} z_{kg}^2$ and $(1/m_n) \sum_{k=1}^{m_n} z_{kg}^2$ converge to $\sigma_{gg}^{(x)}$ in probability. Hence, it may be reasonable to use this statistic for testing the null hypothesis H_0. Under the null hypothesis H_0, we have the next result, the proof of which is given in Chap. 5.

Corollary 3.1 *Assume $0 < \alpha < \beta < 1$ and the conditions of Theorem 3.1. Under $H_0 : \sigma_{gg}^{(v)} = 0$ for some g $(1 \le g \le p)$,*

$$T_1 \xrightarrow{d} N(0, 2) \tag{3.27}$$

as $n \to \infty$.

It is straightforward to construct test statistics and testing procedures based on the SIML estimator, and these are valid asymptotically as the standard statistical procedure. Ait-Sahalia and Xiu (2018) have developed a different testing procedure on the same problem.

3.4   An Optimal Choice of m_n

Because the properties of the SIML estimation method depends crucially on the choice of m_n, which is dependent of n, we have investigated the asymptotic effects as well as the small-sample effects of its choice.

As we will see in Chap. 5 (see the proof of Lemma 5.3), the dominant order of the bias of the SIML estimator is $n^{-1} \sum_{k=1}^{m_n} a_{kn} = O(n^{2\alpha-1})$. Because the normalization of the SIML estimator is in the form of $\sqrt{m_n}[\hat{\sigma}_{gg}^{(x)} - \sigma_{gg}^{(x)}] = O_p(1)$, its variance is of the order $O(n^{-\alpha})$. Hence, when n is large, we can approximate the mean squared error of $\hat{\sigma}_{gg}^{(x)}$ $(g = 1, \ldots, p)$ as

$$g_n(\alpha) = c_{1g} \frac{1}{n^{\alpha}} + c_{2g} n^{4\alpha-2} , \tag{3.28}$$

where c_{1g} and c_{2g} are some constants.

The first and second terms of (3.28) correspond to the order of the variance and the squared bias, respectively. By minimizing $g_n(\alpha)$ with respect to α and using the fact that $\frac{d}{d\alpha}[n^{-\alpha}] = (-\log n)n^{-\alpha}$, we have the condition that $n^{5\alpha-2}$ is constant. Then we can obtain an optimal choice of m_n as follows.

Theorem 3.2 *An optimal choice of* $m_n = [n^\alpha]$ $(0 < \alpha < 0.5)$ *to minimize (3.28) with respect to* α, *when n is large, is given by* $\alpha^* = 0.4$.
When $\alpha = 0.4$

$$\sqrt{m_n}\left[\hat{\sigma}_{gh}^{(x)} - \sigma_{gh}^{(x)} - a\sigma_{gh}^{(v)}\frac{1}{\sqrt{m_n}}\right] \xrightarrow{w} N\left(0, \sigma_{gg}^{(x)}\sigma_{hh}^{(x)} + \left[\sigma_{gh}^{(x)}\right]^2\right), \qquad (3.29)$$

where

$$a = \lim_{n\to\infty}\frac{1}{\sqrt{m_n}}\sum_{k=1}^{n}a_{kn} = \frac{\pi^2}{3}. \qquad (3.30)$$

See the proof of Lemma 5.3 in Chap. 5 for the derivation of a. In this case, we have some asymptotic bias, which is dependent upon the covariance $\sigma_{gh}^{(v)}$.

It is possible to generalize the rule $m_n = [d\, n^\alpha]$, and d is a positive constant. When $p = q = 1$, for instance, we use the notation $\boldsymbol{\Sigma}_x = \sigma_x^2$ and $\boldsymbol{\Sigma}_v = \sigma_v^2$. In this case, we find that $c_{1g} = 2\sigma_x^4 d$ and $c_{2g} = (\pi^2/3)^2\sigma_v^4 d^4$ by ignoring the fractional term of $[\cdot]$. Then we have the conditions that $n^{5\alpha-2}$ is constant and $2\sigma_x^4 d^{-1} = 4(\pi^2/3)^2\sigma_v^4 d^4$. They give the asymptotically optimal choice of α and d as $\alpha^* = 0.4$ and

$$d^* = \left[\frac{9}{2\pi^4}\frac{\sigma_x^4}{\sigma_v^4}\right]^{1/5} \sim 0.541\left(\frac{\sigma_x^2}{\sigma_v^2}\right)^{0.4}. \qquad (3.31)$$

In most cases of our simulations, we obtain reasonable estimates when we set $\alpha = 0.4$ and $d = 1$. It may be problematic to use an estimate of the unknown signal-to-noise ratio with d except $d = 1$ in practical applications. For l_n, we have only the condition $0 < \beta < 1$ and we obtain a reasonable estimate when we set $\beta = 0.8$ by using our results in simulations. There could be some improvements on the finite-sample properties when we use different criteria for choosing m_n.

3.5 Asymptotic Properties of the SIML Estimator When Instantaneous Volatility is Time Varying

It is important to investigate the asymptotic properties of the SIML estimator when the instantaneous volatility function $\boldsymbol{\Sigma}_x(s)$ $(= (\sigma_{gh}^{(x)}(s)))$ of the underlying asset price is not constant over time. When the integrated volatility is a positive (deterministic) constant a.s. (i.e., $\sigma_{gh}^{(x)} = \int_0^1 \sigma_{gh}^{(x)}(s)ds$ is not stochastic) while the instantaneous covariance function is time varying, we have consistency and asymptotic normality for the SIML estimator as $n \to \infty$. We summarize the asymptotic properties of the SIML estimator, the proof of which is again given in Chap. 5.

Theorem 3.3 *We assume that* $\mathbf{x}_i$ *and* $\mathbf{v}_i$ $(i = 1, \ldots, n)$ *in (3.1) and (3.2) are independent and that* $\boldsymbol{\Sigma}_x(s) = \mathbf{C}_x(s)\mathbf{C}_x'(s) \geq 0$, *whose elements are continuous*

and bounded functions. Assume (3.3) and $\boldsymbol{\Sigma}_x$ *is a deterministic matrix and* $\mathbf{v}_i$
are a sequence of independent random vectors with $\mathbf{E}[v_{ig}^2 v_{jh}^2] < \infty$ $(i, j = 1, \ldots,$
$n; g, h = 1, \ldots, p)$.
Define the SIML estimator $\hat{\boldsymbol{\Sigma}}_x = (\hat{\sigma}_{gh}^{(x)})$ *of* $\boldsymbol{\Sigma}_x = (\sigma_{gh}^{(x)})$ *by (3.18) and (3.20), respectively.*
(i) For $m_n = [n^\alpha]$ *and* $0 < \alpha < 0.5$, *as* $n \longrightarrow \infty$

$$\hat{\boldsymbol{\Sigma}}_x - \boldsymbol{\Sigma}_x \xrightarrow{p} \mathbf{O} \,. \tag{3.32}$$

(ii) For $m_n = [n^\alpha]$ *and* $0 < \alpha < 0.4$, *as* $n \longrightarrow \infty$

$$\sqrt{m_n}\left[\hat{\sigma}_{gh}^{(x)} - \sigma_{gh}^{(x)}\right] \xrightarrow{d} N\left[0, V_{gh}\right] \,, \tag{3.33}$$

where

$$V_{gh} = \int_0^1 \left[\sigma_{gg}^{(x)}(s)\sigma_{hh}^{(x)}(s) + \sigma_{gh}^{(x)2}(s)\right] ds \,. \tag{3.34}$$

For the basic case, the above result reduces to Theorem 3.1. We allow the volatility and co-volatility to be deterministic and time varying. Furthermore, we allow $\boldsymbol{\Sigma}_x$ to be a random matrix, but then we need the concept of stable convergence which has been explained by Hall and Heyde (1980), Jacod and Protter (2012), and Häusler and Luschgy (2015). For this purpose, we need to extend the probability space $(\Omega, \mathscr{F}, P)$ to the nice extended probability space $(\tilde{\Omega}, \tilde{\mathscr{F}}, \tilde{P})$. Then we say that a sequence of random variables Z_n with an index n converges stably in law if $\mathbf{E}[Yf(Z_n)] \longrightarrow \tilde{\mathbf{E}}[Yf(Z)]$ for all bounded continuous functions $f(\cdot)$ and all bounded random variables Y, and $\tilde{\mathbf{E}}[\cdot]$ is the expectation operator with respect to the extended probability space. We denote this convergence as $Z_n \xrightarrow{\mathscr{L}-s} Z$. Also we write

$$Z_n \xrightarrow{\mathscr{L}-s} N(0, 2\int_0^1 \sigma_x^4(s)ds) \tag{3.35}$$

if Z_n is a sequence of one-dimensional normalized processes as $\sqrt{m_n}[\hat{\sigma}_x^2 - \sigma_x^2]$ $(p = q = 1)$ and Z is a continuous process defined on a very good filtered extension of $(\Omega, \mathscr{F}, P)$, and conditionally on the sub-σ-field $\mathscr{G}$ of σ-field $\tilde{\mathscr{F}}$, it is a Gaussian process with independent increments satisfying

$$V_x = \tilde{\mathbf{E}}[Z^2|\mathscr{G}] = 2\int_0^1 \sigma_x^4(s)ds \,, \tag{3.36}$$

where the σ-field $\mathscr{G} \subset \tilde{\mathscr{F}}$.

The results of Theorem 3.3 hold in the stochastic case because we can apply the stable convergence results given by Chap. 9 of Jacod and Shiryaev (2003) or Chap. 2 of Jacod and Protter (2012) to our present formulation. The only modification is that the limiting distribution is a mixed-Gaussian distribution.

We consider the general p-dimensional Itô's semi-martingale $X(t)$ with

$$\mathbf{X}(t) = \mathbf{X}(0) + \int_0^t \boldsymbol{\mu}_x(s)ds + \int_0^t \mathbf{C}_x(s)d\mathbf{B}(s) \quad (0 \leq t \leq 1), \tag{3.37}$$

where $\boldsymbol{\mu}_x(s)$ and $\mathbf{C}_x(s)$ are the $p \times 1$ drift terms and the $p \times q$ volatility matrix, respectively, which are bounded and progressively measurable with respect to the σ-field $\mathscr{F}_t$ and $\mathbf{B}(s)$ is q-dimensional Brownian motion $(q \geq 1)$.

In this case, we also make a simple assumption on the (stochastic) volatility function, which follows an Itô's Brownian semi-martingale given by

$$c_{ij}^{(x)}(t) = c_{ij}^{(x)}(0) + \int_0^t \mu_{ij}^\sigma(s)ds + \int_0^t \omega_{ij}^\sigma d\mathbf{B}^\sigma(s), \tag{3.38}$$

where $\mathbf{C}_x(t) = (c_{ij}^{(x)}(t))$ is a $p \times q$ volatility process, $\mathbf{B}^\sigma(s)$ is a $q^* \times 1$ second Brownian motions (which can be correlated with $\mathbf{B}(s)$), $\mu_{ij}^\sigma(s)$ are the drift terms of volatilities, and $\omega_{ij}^\sigma(s)$ $(1 \times q^*)$ are the diffusion terms of instantaneous volatilities, respectively. They are predictable and progressively measurable with respect to $(\Omega, \mathscr{F}, (\mathscr{F}_t)_{t\geq 0}, P)$, and they are bounded and Lipschitz continuous such that the volatility and co-volatility processes are smooth.

We summarize the asymptotic properties of the SIML estimator in the general diffusion case, the proof of which is again given in Chap. 5.

Theorem 3.4 *We assume that* $\mathbf{x}_i$ *and* $\mathbf{v}_i$ $(i = 1, \ldots, n)$ *in (3.1) and (3.37) are independent,* $\boldsymbol{\Sigma}_x(s) = \mathbf{C}_x(s)\mathbf{C}_x'(s) \geq 0$ *and* $\mathbf{C}_x(s)$ *follows (3.38) (* $\boldsymbol{\Sigma}_x$ *is a random matrix) and we have other conditions of Theorem 3.3. We also assume that* $\mu_{ij}^\sigma(s)$ *and* $\omega_{ij}^\sigma(s)$ *are progressively measurable, continuous, and bounded functions. Then the asymptotic results in Theorem 3.3 hold in the sense of stable convergence.*

3.6 Discussion

Although we have introduced the SIML estimator as a modification of the ML estimator in the basic case, Theorems 3.1, 3.3, and 3.4 show that the SIML estimator is consistent and converges weakly to the normal or mixed-normal distribution under more general conditions. Furthermore, it can be shown that the asymptotic properties of the SIML estimator essentially remain the same even when the noise terms are weakly dependent and can be correlated with the signal terms. In the SIML approach, we can separate the information about the covariance matrix of the underlying price volatilities and the covariance matrix of the micro-market noise in an asymptotic sense. Then the resulting estimators of integrated volatility and covariances do not require the assumption of independence among $\mathbf{x}_i$ (the state vector) and $\mathbf{v}_i$ (the noise vector).

Although there are merits in SIML estimation, however, there could naturally be some cost. The convergence rate of the SIML estimator of Σ_x in Theorem 3.1 is slightly less than 0.25 if we take $\alpha = 0.4$ $(\alpha/2 = 0.2)$. Hence, the SIML estimation sacrifices a small efficiency loss against ML estimation based on the MA(1) process when the standard assumptions hold without any misspecification. This is because we have pursued simplicity and applicability of the estimation method, and asymptotic robustness of the procedure for multivariate high-frequency data with possible misspecification. As we will explain in Chaps. 6 and 7, the SIML estimator is asymptotically robust under a variety of practical situations.

References

Ait-Sahalia, Y., and D. Xiu. 2018. A Hausman test for the presence of market microstructure noise in high-frequency data, forthcoming in Journal of Econometrics.

Anderson, T.W. 2003. *An Introduction to Statistical Multivariate Analysis*, 3rd ed. New York: Wiley.

Gloter, A., and J. Jacod. 2001. Diffusions with measurement errors. II: Optimal estimators. *ESAIM: Probability and Statistics* 5: 243–260.

Hall, P., and C. Heyde. 1980. *Martingale Limit Theory and its Applications*. New York: Academic Press.

Hausler, E., and H. Luschgy. 2015. *Stable Convergence and Stable Limit Theorems*. Cham: Springer.

Jacod, J., and P. Protter. 2012. *Discretization of Processes*. Heidelberg: Springer.

Jacod, J., and A.N. Shiryaev. 2003. *Limit Theorems for Stochastic Processes*. Berlin: Springer.

Chapter 4
An Application to Nikkei-225 Futures and Some Simulation

Abstract We present an application of the SIML estimation. We used the high-frequency financial data of the Nikkei-225 Futures, which are the major financial products that are traded actively in Japan. We also give the simulation results of SIML estimation in the basic case and consider the hedging problem, which was the original motivation of developing the SIML method.

4.1 Introduction

An important financial futures market on Nikkei-225 in Japan began started in September 1987 at the Osaka Securities Exchange (OSE), which was the second largest securities exchange after the Tokyo Securities Exchange (TSE) in Japan. Since then, the futures market (not in TSE, but in OSE) has grown in trading size and scale. Nikkei-225 Futures, the most successful products of the OSE, correspond to the Nikkei-225 Spot-Index as its future contracts and the Nikkei-225 spot rate has been the most important stock index in the Japanese financial sector. The trading volume of Nikkei-225 Futures at the OSE has been heavy, and on most days, trades usually occur within one second. The Nikkei-225 Futures have been the major financial tool for risk managements in the financial industry because the Nikkei-225 spot rate has been the major financial index in Japan. We have high-frequency data consisting of prices within less than 1s of Nikkei-225 Futures trades in most time. In our analysis, we have used a set of data with intervals of 1, 5, 10, 30, 60, and 120 s. The SIML estimate of σ_x^2 was calculated by (3.18) with $p = 1$ and $\alpha = .4$.

We chose April 16, 2007, as a typical day before the Lehman Shock in 2008, and we used the same data set of observations. In Table 4.1, we list our estimates of integrated volatility with different time intervals according to both traditional realized volatility (or historical volatility (HI)) estimation and SIML estimation as a typical example. We chose that one day because it was not associated with any significant moves in the OSE and TSE markets and exhibited no significant daily, weekly, monthly, quarterly, or fiscal-year effects, and there were no major and abrupt economic shocks around that day. The values of the estimated HI depend heavily on the observation intervals; for instance, the 1 s volatility estimate is more than 10 times the estimate with the

© The Author(s) 2018

N. Kunitomo et al., *Separating Information Maximum Likelihood Method for High-Frequency Financial Data*, JSS Research Series in Statistics, https://doi.org/10.1007/978-4-431-55930-6_4

Table 4.1 Estimation of integrated volatility

	$\hat{\sigma}_x^2$	RV (HI)
1 s	4.085E-05	4.946E-04
5 s	3.994E-05	2.601E-04
10 s	4.990E-05	1.764E-04
30 s	3.551E-05	9.449E-05
60 s	4.550E-05	6.964E-05
120 s	4.750E-05	6.057E-05

60 s. This phenomenon regarding the calculated RV has been observed in many high-frequency financial data and has been one of the main reasons why we need to take account of micro-market noise as we have mentioned in Sect. 2.3. Several researchers have highlighted the problem of significant biases in the estimated HI or RV, and our analysis has been consistent with them. Also, by using the test statistic in (3.26) and (3.27), we find that $T_1 = 103.56(1\,s)$, $43.26(5\,s)$, $19.15(10\,s)$, $11.29(30\,s)$, and $3.07(60\,s)$. Most of these values are highly significant. This statistic also shows that when we have longer time intervals as more than one minute, the effects of micro-market noise become relatively small.

By contrast, the estimates based on the SIML method are more stable over the different sampling intervals and may be reasonable as in Table 4.1. There are some variations, but they are within the range of sampling effects. Thus, we have confirmed that the presence of micro-market noise is an important factor when we have high-frequency data in the Nikkei-225 Futures market.

The analysis of the Nikkei-225 spot and futures markets with bivariate high-frequency data was the real motivation for developing our approach in this book, and we will illustrate this hedging problem in this chapter.

There can be other empirical applications of the SIML approach, and Misaki (2018) has given one example.

4.2   Basic Simulation Results

We have also conducted a large number of simulations and, we first present three representative cases among many possibilities. We chose the sample sizes $n = 300$, 5,000, and 20,000, which correspond roughly to data intervals 1 min, 10 and 1 s, respectively. We also report one simulation of hedging in Table 4.6, but we omit the result simulations on the cases of stochastic volatility, autocorrelated noise, and endogenous noise. The number of replications in our simulations is basically 5,000. We conducted a two-dimensional simulation for Table 4.6 because we need two simulated data sets for the spot price series and the futures price series. The other tables are based on one-dimensional simulations because we need only one simulated

Table 4.2 Estimated
noise-variance

	$\hat{\sigma}_v^2$
5 s	3.009E-08
10 s	4.168E-08
60 s	8.976E-08
120 s	6.834E-08

data set for the futures price series, and we have utilized the formulation of the estimation problem in Chap. 3 when $p = q = 1$.

In simulation, it is important to set the reasonable value of signal and noise variances. For this purpose, we calculated the SIML estimate of noise variance by using the same data set, which is given in Table 4.2. The SIML estimate of σ_v^2 was calculated by (3.20) with $p = q = 1$ and $\beta = .8$. We have found that the variance of noise is about $10^{-2} \sim 10^{-3}$ of the variance or volatility level in this case, which gives a standard value for later simulations.

In our simulations, Tables 4.3, 4.4, and 4.5 give the SIML estimation results for integrated volatility, where $\sigma_{ff}^{(v)}$ stands for the variance of noise and H-vol stands for the historical volatility (i.e., realized volatility). Table 4.3 corresponds to the standard case of flat volatility, while Table 4.4 corresponds to the case of a U-shaped intraday volatility function. The integrated volatility and the (instantaneous) volatility function are defined by

$$\sigma_{ff}^{(x)} = \int_0^1 \sigma_{ff}^{(x)}(s)ds \qquad (4.1)$$

and

$$\sigma_{ff}^{(x)}(s) = \sigma_{ff}^{(x)}(0)\left[a_0 + a_1 s + a_2 s^2\right] \ (0 \le s \le 1), \qquad (4.2)$$

respectively, where a_i $(i = 1, 2, 3)$ are constants satisfying $\sigma_{ff}^{(x)}(s) > 0$ for $s \in [0, 1]$. The flat volatility function of Table 4.3 corresponds to the case with $a_1 = a_2 = 0$.

First, when the noise variance is small and $n = 300$ (the interval length is several minutes), the H-vol does not differ from the true value substantially. Even when the noise level is small, however, the bias of H-vol becomes large for $n = 5,000$ and $n = 20,000$. In most cases, the H-vol is completely unreliable.

Second, the SIML estimates in Tables 4.3 and 4.4 are reliable, and their biases are of smaller order.

Third, the SD (standard deviation) and MSE are reasonable and the AVAR (the asymptotic variance given in Theorem 3.1 of Chap.3) are useful approximations in all cases.

Fourth, Table 4.5 gives the SIML estimation result when the signal term x_{fi} and the noise term v_f follow t-distributions. Strictly speaking, this case does not correspond to the model of (4.1) and (4.2) with a deterministic volatility function. In some stochastic volatility cases, the signal term may follow a t-distribution, but

Table 4.3 Estimation of realized volatility (constant volatility, $\alpha = 0.4$)

n = 300	$\sigma_{ff}^{(x)}$	$\sigma_{ff}^{(v)}$	H-vol	$\sigma_{ff}^{(x)}$	$\sigma_{ff}^{(v)}$	H-vol	$\sigma_{ff}^{(x)}$	$\sigma_{ff}^{(v)}$	H-vol
True	2.00E-04	2.00E-06		2.00E-04	2.00E-07		2.00E-04	2.00E-09	
Mean	2.05E-04	2.17E-06	1.40E-03	2.03E-04	3.80E-07	3.20E-04	2.01E-04	1.84E-07	2.01E-04
SD	9.75E-05	3.21E-07	1.34E-04	9.71E-05	5.51E-08	2.74E-05	9.42E-05	2.64E-08	1.64E-05
MSE	9.53E-09	1.33E-13		9.44E-09	3.55E-14		8.88E-09	3.37E-14	
AVAR	8.17E-09	8.34E-14		8.17E-09	8.34E-16		8.17E-09	8.34E-20	
n = 5,000	$\sigma_{ff}^{(x)}$	$\sigma_{ff}^{(v)}$	H-vol	$\sigma_{ff}^{(x)}$	$\sigma_{ff}^{(v)}$	H-vol	$\sigma_{ff}^{(x)}$	$\sigma_{ff}^{(v)}$	H-vol
True	2.00E-04	2.00E-06		2.00E-04	2.00E-07		2.00E-04	2.00E-09	
Mean	2.06E-04	2.01E-06	2.02E-02	2.01E-04	2.10E-07	2.20E-03	2.01E-04	1.23E-08	2.20E-04
SD	5.37E-05	9.44E-08	4.93E-04	5.24E-05	9.74E-09	5.18E-05	5.21E-05	5.82E-10	4.43E-06
MSE	2.91E-09	8.98E-15		2.75E-09	1.98E-16		2.72E-09	1.06E-16	
AVAR	2.65E-09	8.79E-15		2.65E-09	8.79E-17		2.65E-09	8.79E-21	
n = 20,000	$\sigma_{ff}^{(x)}$	$\sigma_{ff}^{(v)}$	H-vol	$\sigma_{ff}^{(x)}$	$\sigma_{ff}^{(v)}$	H-vol	$\sigma_{ff}^{(x)}$	$\sigma_{ff}^{(v)}$	H-vol
True	2.00E-04	2.00E-06		2.00E-04	2.00E-07		2.00E-04	2.00E-09	
Mean	2.06E-04	2.00E-06	8.02E-02	2.00E-04	2.03E-07	8.20E-03	2.00E-04	4.54E-09	2.80E-04
SD	4.03E-05	5.24E-08	9.78E-04	3.95E-05	5.43E-09	1.01E-04	3.90E-05	1.24E-10	2.85E-06
MSE	1.66E-09	2.76E-15		1.56E-09	3.65E-17		1.52E-09	6.46E-18	
AVAR	1.52E-09	2.90E-15		1.52E-09	2.90E-17		1.52E-09	2.90E-21	

Note: the data generating process is

$y_{fi} = x_{fi} + v_{fi}$, $x_{fi} = x_{f,i-1} + u_i$,

where $u_i \sim i.i.d.N(0, \sigma_{ff}^{(x)}(1/m))$, $v_{fi} \sim i.i.d.N(0, \sigma_{ff}^{(v)})$

Table 4.4 Estimation of realized volatility (U-shaped volatility, $\alpha = 0.4$)

n = 300	$\sigma_{ff}^{(x)}$	$\sigma_{ff}^{(v)}$	H-vol	$\sigma_{ff}^{(x)}$	$\sigma_{ff}^{(v)}$	H-vol	$\sigma_{ff}^{(x)}$	$\sigma_{ff}^{(v)}$	H-vol
True	1.67E-04	2.00E-06		1.67E-04	2.00E-07		1.67E-04	2.00E-09	
Mean	1.71E-04	2.15E-06	1.37E-03	1.68E-04	3.53E-07	2.88E-04	1.68E-04	1.53E-07	1.68E-04
SD	8.17E-05	3.18E-07	1.33E-04	7.91E-05	5.14E-08	2.45E-05	7.96E-05	2.26E-08	1.39E-05
MSE	6.68E-09	1.23E-13		6.25E-09	2.59E-14		6.33E-09	2.34E-14	
AVAR	5.67E-09	8.34E-14		5.67E-09	8.34E-16		5.67E-09	8.34E-20	
n = 5,000	$\sigma_{ff}^{(x)}$	$\sigma_{ff}^{(v)}$	H-vol	$\sigma_{ff}^{(x)}$	$\sigma_{ff}^{(v)}$	H-vol	$\sigma_{ff}^{(x)}$	$\sigma_{ff}^{(v)}$	H-vol
True	1.67E-04	2.00E-06		1.67E-04	2.00E-07		1.67E-04	2.00E-09	
Mean	1.72E-04	2.01E-06	2.02E-02	1.67E-04	2.09E-07	2.17E-03	1.67E-04	1.06E-08	1.87E-04
SD	4.52E-05	9.34E-08	4.96E-04	4.27E-05	9.87E-09	5.22E-05	4.30E-05	4.94E-10	3.76E-06
MSE	2.07E-09	8.76E-15		1.82E-09	1.75E-16		1.85E-09	7.38E-17	
AVAR	1.84E-09	8.79E-15		1.84E-09	8.79E-17		1.84E-09	8.79E-21	
n = 20,000	$\sigma_{ff}^{(x)}$	$\sigma_{ff}^{(v)}$	H-vol	$\sigma_{ff}^{(x)}$	$\sigma_{ff}^{(v)}$	H-vol	$\sigma_{ff}^{(x)}$	$\sigma_{ff}^{(v)}$	H-vol
True	1.67E-04	2.00E-06		1.67E-04	2.00E-07		1.67E-04	2.00E-09	
Mean	1.71E-04	2.00E-06	8.02E-02	1.68E-04	2.02E-07	8.17E-03	1.67E-04	4.12E-09	2.47E-04
SD	3.47E-05	5.43E-08	9.68E-04	3.32E-05	5.39E-09	9.86E-05	3.31E-05	1.12E-10	2.54E-06
MSE	1.22E-09	2.95E-15		1.11E-09	3.34E-17		1.09E-09	4.49E-18	
AVAR	1.06E-09	2.90E-15		1.06E-09	2.90E-17		1.06E-09	2.90E-21	

Note: the data generating process is

$y_{fi} = x_{fi} + v_{fi}, x_{fi} = x_{f,i-1} + u_i$

$u_i \sim i.i.d. N(0, (1-s+s^2)\sigma_{ff}^{(x)}(0)(1/n)), v_{fi} \sim i.i.d. N(0, \sigma_{ff}^{(v)}), s = i/n \ (i = 1, \ldots, n)$

Table 4.5 T-error distribution ($df_x = 5, df_v = 7, \alpha = 0.4$)

n = 300	$\sigma_{ff}^{(x)}$	$\sigma_{ff}^{(v)}$	H-vol	$\sigma_{ff}^{(x)}$	$\sigma_{ff}^{(v)}$	H-vol	$\sigma_{ff}^{(x)}$	$\sigma_{ff}^{(x)}$	H-vol
True	8.33E-05	7.00E-07		8.33E-05	7.00E-08		8.33E-05	7.00E-10	
Mean	8.54E-05	7.79E-07	5.06E-04	8.30E-05	1.46E-07	1.25E-04	8.40E-05	7.65E-08	8.37E-05
SD	4.23E-05	1.27E-07	5.96E-05	4.10E-05	2.38E-08	1.53E-05	4.16E-05	1.48E-08	1.29E-05
MSE	1.79E-09	2.22E-14		1.68E-09	6.29E-15		1.73E-09	5.96E-15	
AVAR	1.42E-09	1.02E-14		1.42E-09	1.02E-16		1.42E-09	1.02E-20	
n = 5,000	$\sigma_{ff}^{(x)}$	$\sigma_{ff}^{(v)}$	H-vol	$\sigma_{ff}^{(x)}$	$\sigma_{ff}^{(v)}$	H-vol	$\sigma_{ff}^{(x)}$	$\sigma_{ff}^{(x)}$	H-vol
True	8.33E-05	7.00E-07		8.33E-05	7.00E-08		8.33E-05	7.00E-10	
Mean	8.54E-05	7.04E-07	7.08E-03	8.32E-05	7.42E-08	7.83E-04	8.33E-05	4.99E-09	9.03E-05
SD	2.24E-05	3.57E-08	2.21E-04	2.16E-05	3.77E-09	2.36E-05	2.14E-05	2.73E-10	3.38E-06
MSE	5.06E-10	1.30E-15		4.67E-10	3.22E-17		4.59E-10	1.84E-17	
AVAR	4.60E-10	1.08E-15		4.60E-10	1.08E-17		4.60E-10	1.08E-21	
n = 20,000	$\sigma_{ff}^{(x)}$	$\sigma_{ff}^{(v)}$	H-vol	$\sigma_{ff}^{(x)}$	$\sigma_{ff}^{(v)}$	H-vol	$\sigma_{ff}^{(x)}$	$\sigma_{ff}^{(x)}$	H-vol
True	8.33E-05	7.00E-07		8.33E-05	7.00E-08		8.33E-05	7.00E-10	
Mean	8.51E-05	7.01E-07	2.81E-02	8.33E-05	7.10E-08	2.88E-03	8.32E-05	1.76E-09	1.11E-04
SD	1.70E-05	2.01E-08	4.42E-04	1.64E-05	2.06E-09	4.47E-05	1.66E-05	5.22E-11	1.80E-06
MSE	2.91E-10	4.05E-16		2.69E-10	5.33E-18		2.74E-10	1.12E-18	
AVAR	2.64E-10	3.55E-16		2.64E-10	3.55E-18		2.64E-10	3.55E-22	

Note: the data generating process is

$$y_{fi} = x_{fi} + \sqrt{\sigma_{ff}^{(v)}}\, v_{fi}$$

$$x_{fi} = x_{fi-1} + \sqrt{\sigma_{ff}^{(x)}(1/n)}\, u_t,$$

$u_i \sim i.i.d.t(df_x), v_t \sim i.i.d.t(df_v),$

where df_x and df_v are the degrees of freedom of t(df) and the variance of u_i is normalized such that the resulting volatility is comparable to other cases

Fig. 4.1 Nikkei-225 spot
and futures (1 min frequency
data)

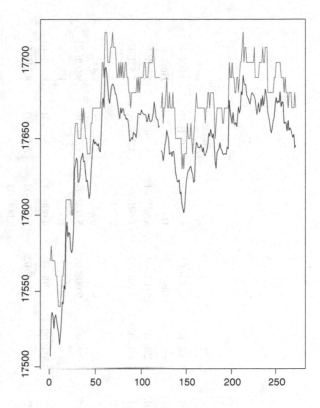

those situations are far more complicated than the present discussion, and we do
not pursue the resulting problems herein. Nonetheless, this simulation result may be
interesting because it shows that the SIML estimation is not overly dependent on
the assumption of Gaussian distribution. Since the SIML estimate gives the similar
results as Tables 4.3 and 4.4, we confirm that it is robust against the non-normality
for both the noise term and diffusion process.

To summarize the findings from our basic simulations, the SIML estimation gives
stable and reasonable estimates even when the (instantaneous) volatility function
changes during a day.

4.3 Realized Hedging

An important use of realized volatility and realized co-volatility (or covariance) is to
form the (estimated) realized hedging coefficient. The use of financial futures as the
risk hedging instruments for spot-securities is quite common in financial industries,
and the details of risk hedging problem are explained by Duffie (1989) for instance.
In this respect, Nikkei-225 futures is the major hedging tool to the Nikkei-225 spot

Table 4.6 Hedge ratio ($\alpha = 0.4$, $corv = 0$)

n = 300	$\sigma_{ff}^{(x)}$	$\sigma_{ff}^{(v)}$	corx	$\sigma_{ff}^{(x)}$	$\sigma_{ff}^{(v)}$	corx	$\sigma_{ff}^{(x)}$	$\sigma_{ff}^{(v)}$	corx
True	5.00E-05	5.00E-07	9.00E-01	5.00E-05	5.00E-08	9.00E-01	5.00E-05	5.00E-10	9.00E-01
	Hx	corv	Hh	Hx	corv	Hh	Hx	corv	Hh
Mean	8.71E-01	7.63E-02	1.29E-01	9.01E-01	4.29E-01	5.63E-01	8.98E-01	8.90E-01	8.94E-01
SD	1.85E-01	1.03E-01	6.84E-02	1.69E-01	9.47E-02	5.20E-02	1.64E-01	4.77E-02	2.57E-02
MSE	3.49E-02			2.87E-02			2.71E-02		
n = 5,000	$\sigma_{ff}^{(x)}$	$\sigma_{ff}^{(v)}$	corx	$\sigma_{ff}^{(x)}$	$\sigma_{ff}^{(v)}$	corx	$\sigma_{ff}^{(x)}$	$\sigma_{ff}^{(v)}$	corx
True	5.00E-05	5.00E-07	9.00E-01	5.00E-05	5.00E-08	9.00E-01	5.00E-05	5.00E-10	9.00E-01
	Hx	corv	Hh	Hx	corv	Hh	Hx	corv	Hh
Mean	8.78E-01	5.21E-03	9.61E-03	8.97E-01	4.35E-02	8.20E-02	9.00E-01	7.54E-01	8.18E-01
SD	9.50E-02	3.32E-02	1.71E-02	8.29E-02	3.31E-02	1.70E-02	8.37E-02	2.18E-02	8.23E-03
MSE	9.50E-03			6.89E-03			7.00E-03		

Note: the data generating process is

$y_{j,i} = x_{j,i} + v_{j,i}$ $(j = s, f)$, $x_{j,i} = x_{j,i-1} + u_{j,i}$,

$u_{j,i} \sim i.i.d.N(0, \sigma_{jj}^{(x)}(1/n))$, $v_{j,i} \sim i.i.d.N(0, \sigma_{jj}^{(v)})$, $corr(u_{s,i}, u_{f,i}) = corx$, $corr(v_{s,i}, v_{f,i}) = corv$

rate (which is the major stock index of Japan) and other stock prices in the Japanese financial industries. It is because by using Nikkei-225 futures, it may be possible to replicate the Nikkei-225 spot index quite easily. For an illustration, we show the Nikkei-225 futures and spot-rate in Fig. 4.1 within in one day. (Colored one means future data, while Black means spot data.)

We have also conducted number of simulations of hedging. Table 4.6 gives the SIML estimation simulation on spot-rate, futures-rate, and their covariances for the hedging problem, where H_x stands for the SIML estimate, while Hh stands for the estimate based on the historical volatility and covariance. When the noise variance is small, the value of hedging ratio based on realized volatility and co-volatility looks reasonable. When the noise is not small, however, its bias becomes significant. The SIML estimates of the covariance for the hedging ratio also give stable and reasonable estimates in most cases.

Based on the volatility and covariance estimates, we have replicated the SIML estimate of the hedging ratio in number of different situations. Unlike with other methods, we find that our estimates are stable and reliable. The estimated values of RV and its hedging estimates (HI) vary from day to day and often deviate significantly different from 1 when we use high-frequency data and the volatility function is time-varying. By contrast, the SIML estimated values are often close to unity, which agrees with the intuitive reasoning among the market participants. The most important finding is that the hedging ratio estimate from high-frequency historical data is unreliable, whereas those from the SIML estimation are reasonable. Then, it may be possible that by using the high-frequency data, each day we can estimate the hedging strategy with the optimal hedging ratio and then determine the hedging strategy for the next day. When we use only historical data and RV, we hedged risk poorly. However, we have found that the SIML-based hedging performs reasonably well in the sense that it approximates the pure futures hedging.

We have conducted a large number of simulations for the hedging problem because it is important for financial risk managements and some of them will be reported in Chap. 7 under a variety of different situation of micro-markets.

References

Duffie, D. 1989. *Futures Markets*. Estados Unidos: Prentice Hall.
Misaki, H. 2018. An empirical analysis of volatility by the SIML estimation with high-frequency trades and quotes, Unpublished Manuscript.

Chapter 5
Mathematical Derivations

Abstract We give the mathematical derivations of the results on the asymptotic properties of the SIML estimator in Chap. 3. For the sake of completeness, we also give useful relations of trigonometric functions that are the results of direct but often tedious calculations. This chapter can be skipped for readers who are interested in only financial applications.

5.1 Some Lemmas

Proof of Lemma 3.1: (i) Let $\mathbf{A}_n = (a_{ij})$ in (3.6) $(i, j = 1, \ldots, n)$ and an $n \times 1$ vector $\mathbf{x} = (x_t)$ $(t = 1, \ldots, n)$ satisfying $\mathbf{A}_n \mathbf{x} = \lambda \mathbf{x}$. Then

$$\frac{x_1 + x_2}{2} = \lambda x_1 , \tag{5.1}$$

$$\frac{x_{t-1} + x_{t+1}}{2} = \lambda x_t \ (t = 2, \ldots, n - 1) , \tag{5.2}$$

$$\frac{1}{2} x_{n-1} = \lambda x_n . \tag{5.3}$$

Let ξ_i $(i = 1, 2)$ be the solutions of $\xi^2 - 2\lambda\xi + 1 = 0$. Because $2\lambda = \xi_1 + \xi_2$ and $\xi_1 \xi_2 = 1$, we have the solution of (5.2) as $x_t = c_1 \xi_1^t + c_2 \xi_1^{-t}$ $(t = 1, \ldots, n)$ and c_i $(i = 1)$ are real constants. Then (5.1) implies

$$\begin{aligned} 0 &= c_1 \xi_1 + c_2 \xi_1^{-1} + c_1 \xi_1^2 + c_2 \xi_1^{-2} - (\xi_1 + \xi_1^{-1})(c_1 \xi_1 + c_2 \xi_1^{-1}) \\ &= (\xi_1 - 1)(c_1 - c_2 \xi_1^{-1}) . \end{aligned}$$

Because $\xi = 1$ cannot be a solution (i.e. it leads to a contradiction), we find that $c_2 = c_1 \xi_1$. Then we find that $x_t = c_1 [\xi_1^t + \xi_1^{-(t-1)}]$ and (5.3) implies $\xi_1^{2n+1} = -1$ and

$$\lambda_k = \cos\left[\pi \frac{2k - 1}{2n + 1}\right] \ (k = 1, \ldots, n) . \tag{5.4}$$

© The Author(s) 2018
N. Kunitomo et al., *Separating Information Maximum Likelihood Method for High-Frequency Financial Data*, JSS Research Series in Statistics, https://doi.org/10.1007/978-4-431-55930-6_5

By taking $c_1 = (1/2)\xi_1^{-1/2}$, the elements of the characteristic vectors of $\mathbf{A}_n$ with $\cos[\pi(2k-1)/(2n+1)]$ are

$$x_t = \frac{1}{2}\left[\xi_1^{t-1/2} + \xi_1^{-(t-1/2)}\right] = \cos\left[\pi\frac{2k-1}{2n+1}\left(t - \frac{1}{2}\right)\right]. \tag{5.5}$$

(ii) The rest of the proof involves the standard arguments of spectral decomposition in linear algebra. $\qquad\square$

Lemma 5.1 *(i) For any integers l, m and n $(1 < l, m \le n)$, we have*

$$\sum_{k=1}^{m}\left[\cos\pi\frac{2k-1}{2n+1}l\right] = \frac{1}{2}\frac{\sin 2\pi m\frac{l}{2n+1}}{\sin\pi\frac{l}{2n+1}} \tag{5.6}$$

and

$$\sum_{k=1}^{m}\left[\cos\pi\frac{2k-1}{2n+1}l\right]^2 = \frac{m}{2} + \frac{1}{4}\frac{\sin 4\pi m\frac{l}{2n+1}}{\sin 2\pi\frac{l}{2n+1}}. \tag{5.7}$$

(ii) For any integer k, we have

$$\sum_{t=1}^{n}\left[\cos\pi\frac{2k-1}{2n+1}\left(t - \frac{1}{2}\right)\right]^2 = \frac{n}{2} + \frac{1}{4}. \tag{5.8}$$

Proof of Lemma 5.1: We use the relation

$$\sum_{t=1}^{m}\left(e^{i2\pi\left[\frac{l}{2n+1}\left(t-\frac{1}{2}\right)\right]} + e^{-i2\pi\left[\frac{l}{2n+1}\left(t-\frac{1}{2}\right)\right]}\right)$$

$$= e^{i2\pi\left[\frac{l}{2n+1}\right]\frac{1}{2}} \times \frac{1 - e^{i2\pi\left[\frac{l}{2n+1}\right]m}}{1 - e^{i2\pi\left[\frac{l}{2n+1}\right]}} + e^{-i2\pi\left[\frac{l}{2n+1}\right]\frac{1}{2}} \times \frac{1 - e^{-i2\pi\left[\frac{l}{2n+1}\right]m}}{1 - e^{-i2\pi\left[\frac{l}{2n+1}\right]}}$$

$$= \frac{1}{1 - e^{i2\pi\left[\frac{l}{2n+1}\right]}}\left(e^{i2\pi\left[\frac{l}{2n+1}\right]\frac{1}{2}} - e^{2\pi i\left[\frac{l}{2n+1}\frac{2m+1}{2}\right]} - e^{2\pi i\left[\frac{l}{2n+1}\frac{1}{2}\right]} + e^{-i2\pi\left[\frac{l}{2n+1}\left(\frac{2m-1}{2}\right)\right]}\right)$$

$$= \frac{1}{e^{i\pi\left[\frac{l}{2n+1}\right]} - e^{-i\pi\left[\frac{l}{2n+1}\right]}}\left(e^{i2\pi\left[\frac{l}{2n+1}\right]m} - e^{-i2\pi\left[\frac{l}{2n+1}\right]m}\right).$$

Then we find (5.6). When we take $m = n$ and l is an odd integer, $e^{i\pi l} = -1$ and the above equation equals to

$$\frac{-e^{\pi i\left[\frac{l}{2n+1}(2n+1)\right]} + e^{-\pi\left[\frac{l}{2n+1}(2n+1-2)\right]}}{1 - e^{i2\pi\left[\frac{l}{2n+1}\right]}} = 1.$$

Then by using the relation

$$\sum_{t=1}^{m}\left[\cos\pi\frac{2k-1}{2n+1}\left(t-\frac{1}{2}\right)\right]^2 = \sum_{t=1}^{m}\left[\frac{1}{2}+\frac{1}{2}\cos\pi\frac{2k-1}{2n+1}2\left(t-\frac{1}{2}\right)\right]$$

$$= \frac{m}{2}+\frac{1}{2}\sum_{t=1}^{m}\cos 2\pi\frac{2k-1}{2n+1}\left(t-\frac{1}{2}\right),$$

we have(5.7) and (5.8). $\qquad\square$

Lemma 5.2 *Let* $c_{ij}=(2/m)\sum_{k=1}^{m}s_{ik}s_{jk}$ $(i,j=1,\ldots,n)$ *and*

$$s_{jk}=\cos\left[\frac{2\pi}{2n+1}\left(j-\frac{1}{2}\right)\left(k-\frac{1}{2}\right)\right]. \qquad (5.9)$$

(These c_{ij} are the same as c_{ij}^{*} in Sect. 3.2 except for the constant factor $n/(n+\frac{1}{2})$.)
Then we have
(i) for any integers j, k

$$\sum_{i=1}^{n}c_{ij}c_{ik}=\frac{1}{m}\left(\frac{n}{2}+\frac{1}{4}\right)c_{jk} \qquad (5.10)$$

and

$$\sum_{i,j=1}^{n}c_{ij}^{2}=\frac{4}{m}\left[\frac{n}{2}+\frac{1}{4}\right]^{2}. \qquad (5.11)$$

(ii) As $n\to\infty$,

$$\frac{1}{n}\sum_{i=1}^{n}(c_{ii}-1)\to 0 \qquad (5.12)$$

and

$$\frac{1}{n}\sum_{i=1}^{n}(c_{ii}-1)^{2}\to 0. \qquad (5.13)$$

Proof of Lemma 5.2: We set

$$c_{ij}=\frac{2}{m}\sum_{k=1}^{m}s_{ik}s_{jk}$$

$$=\frac{1}{m}\sum_{k=1}^{m}\left\{\cos\left[\frac{2\pi}{2n+1}(i+j-1)\left(k-\frac{1}{2}\right)\right]+\cos\left[\frac{2\pi}{2n+1}(i-j)\left(k-\frac{1}{2}\right)\right]\right\}.$$

Then by using Lemma 5.1

$$c_{jj} = 1 + \frac{1}{m}\sum_{k=1}^{m}\cos\left[\frac{\pi}{2n+1}(2j-1)(2k-1)\right] \tag{5.14}$$

$$= 1 + \frac{1}{2m}\frac{\sin\left[\frac{2\pi m}{2n+1}(2j-1)\right]}{\sin\left[\frac{\pi}{2n+1}(2j-1)\right]}$$

and

$$\sum_{j=1}^{n}c_{jj} = n + \frac{1}{m}\sum_{k=1}^{m}\left[\sum_{j=1}^{n}\cos\left[\frac{2\pi}{2n+1}(2j-1)\left(k-\frac{1}{2}\right)\right]\right] \tag{5.15}$$

$$= n + \frac{1}{m}\sum_{k=1}^{m}\frac{1}{2}\frac{\sin\left[\frac{2\pi n}{2n+1}(2k-1)\right]}{\sin\left[\frac{\pi}{2n+1}(2k-1)\right]},$$

which is $n + 1/2$ because for $k \geq 1$

$$\sin\left[\frac{\pi(2n+1-1)}{2n+1}(2k-1)\right]$$

$$= \sin[\pi(2k-1)]\cos\left[\frac{\pi}{2n+1}(2k-1)\right] - \cos[\pi(2k-1)]\sin\left[\frac{\pi}{2n+1}(2k-1)\right]$$

$$= \sin\left[\frac{\pi}{2n+1}(2k-1)\right].$$

Similarly,

$$c_{ij}^2 = \left(\frac{2}{m}\right)^2\sum_{k,k'=1}^{m}s_{ik}s_{jk}s_{ik'}s_{jk'}$$

$$= (\frac{1}{m})^2\sum_{k,k'=1}^{m}\left[\cos\left[\frac{2\pi}{2n+1}(i+j-1)\left(k-\frac{1}{2}\right)\right] + \cos\left[\frac{2\pi}{2n+1}(i-j)\left(k-\frac{1}{2}\right)\right]\right]$$

$$\times\left[\cos\left[\frac{2\pi}{2n+1}(i+j-1)\left(k'-\frac{1}{2}\right)\right] + \cos\left[\frac{2\pi}{2n+1}(i-j)\left(k'-\frac{1}{2}\right)\right]\right]$$

$$= (\frac{1}{m})^2\sum_{k,k'=1}^{m}\left\{\frac{1}{2}\cos\left[\frac{2\pi}{2n+1}(i+j-1)(k+k'-1)\right]\right.$$

$$+\frac{1}{2}\cos\left[\frac{2\pi}{2n+1}(i+j-1)(k-k')\right]$$

$$+2\left[\frac{1}{2}\cos\left[\frac{2\pi}{2n+1}((i+j-1)\left(k-\frac{1}{2}\right) + \left(k'-\frac{1}{2}\right)(i-j)\right)\right]$$

$$+\frac{1}{2}\cos\left[\frac{2\pi}{2n+1}\left((i+j-1)\left(k-\frac{1}{2}\right)-(k'-\frac{1}{2})(i-j)\right)\right]$$

$$+\frac{1}{2}\cos\left[\frac{2\pi}{2n+1}(i-j)(k+k'-1)\right]+\frac{1}{2}\cos\left[\frac{2\pi}{2n+1}(i-j)(k-k')\right]\right\}.$$

Hence

$$c_{jj}^2=\left(\frac{2}{m}\right)^2\sum_{k,k'=1,i=j}^{m}s_{ik}s_{jk}s_{ik'}s_{jk'}$$

$$=1+\frac{2}{m}\sum_{k=1}^{m}\cos\left[\frac{2\pi}{2n+1}\left(k-\frac{1}{2}\right)(2j-1)\right]$$

$$+\frac{1}{2m^2}\sum_{k,k'=1}^{m}\left[\cos\left[\frac{2\pi}{2n+1}(2j-1)(k+k'-1)\right]+\cos\left[\frac{2\pi}{2n+1}(2j-1)(k-k')\right]\right].$$

We use the relation $\sum_{j=1}^{n}\cos\left[\frac{2\pi}{2n+1}l(2j-1)\right]=(-1/2)\cos\pi l$, which is $1/2$ for l being an odd integer and $-1/2$ for l being an even integer, and $k\geq 1$. To evaluate $\sum_{j=1}^{n}(c_{jj}^2-1)$, we apply the formula of c_{jj}^2 and we need to calculate the cases in which $\sum_{k=k'}$ and $\sum_{k\neq k'}$, separately. Then we find

$$\frac{1}{n}\sum_{i=1}^{n}(c_{ii}^2-1)=\frac{2}{m}\sum_{k=1}^{m}\frac{1}{n}\sum_{j=1}^{n}s_{jk}+\frac{1}{2m^2}\sum_{k,k'=1}^{m}\left[\frac{1}{n}\sum_{j=1}^{n}s_{j,k+k'-1}\mid\frac{1}{n}\sum_{j=1}^{n}s_{j,k-k'}\right]$$

$$=\frac{2}{m}\sum_{k=1}^{m}\frac{1}{n}\frac{1}{2}+\frac{1}{2m^2}\sum_{k,k'=1}^{m}\frac{1}{n}\left(-\frac{1}{2}\right)+\frac{1}{2m^2}\sum_{k,k'=1}^{m}\left[\frac{1}{n}n\delta(k,k')+\frac{1}{n}\delta(k\neq k')\left(-\frac{1}{2}\right)\right]$$

$$=\frac{1}{2m}+\frac{1}{2n}+\frac{1}{4mn},$$

and then

$$\frac{1}{n}\sum_{i=1}^{n}(c_{ii}-1)^2=\frac{1}{n}\sum_{i=1}^{n}[c_{ii}^2-1-2(c_{ii}-1)]=\frac{1}{2m}+\frac{1}{4mn}-\frac{1}{2n}.$$

Since $\sum_{j=1}^{n}s_{jk}s_{jk'}=\delta_{kk'}(\frac{n}{2}+\frac{1}{4})$,

$$m\sum_{i,j=1}^{n}c_{ij}^2=\frac{4}{m}\sum_{k,k'=1}^{m}\sum_{i,j=1}^{n}s_{ik}s_{ik'}s_{jk}s_{jk'}=\left(n+\frac{1}{2}\right)^2. \qquad (5.16)$$

Here we use the notation the indicator function that $\delta_{kk'}=1$ $(k=k')$ and $\delta_{kk'}=\delta(k\neq k')=0$ $(k\neq k')$.

5.2 Proofs of Theorems

We now prove Theorems 3.1, 3.3 and 3.4 of Chap. 3. Because Theorem 3.1 is essentially a special case of Theorem 3.3 except the latter half of Theorem 3.1, we focus on proving Theorems 3.3 and 3.4. The proof of some parts of Theorem 3.1 will be given as Lemma 5.5 and some additional arguments, which are presented after the proof of Lemma 5.8 below.

We set K_i $(i \geq 1)$ as positive constants in the following derivations. For any unit vector $\mathbf{e}_g = (0, \ldots, 0, 1, 0, \ldots, 0)'$ $(g = 1, \ldots, p)$, we define $\sigma_{gh}^{(x)} = \mathbf{e}_g' \Sigma_x \mathbf{e}_h$, $\hat{\sigma}_{gh}^{(x)} = \mathbf{e}_g' \hat{\Sigma}_x \mathbf{e}_h$, $\sigma_{gh}^{(v)} = \mathbf{e}_g' \Sigma_v \mathbf{e}_h$ and $\hat{\sigma}_{gh}^{(v)} = \mathbf{e}_g' \hat{\Sigma}_v \mathbf{e}_h$. With the transformation of (3.12) we set $z_{kg} = \mathbf{e}_g' \mathbf{z}_k$ $(k = 1, \ldots, n)$ and $z_{kg} = z_{kg}^{(1)} + z_{kg}^{(2)}$, where $z_{kg}^{(1)}$ and $z_{kg}^{(2)}$ correspond to the (k, g)−elements of $\mathbf{Z}_n^{(1)} = h_n^{-1/2} \mathbf{P}_n \mathbf{C}_n^{-1} (\mathbf{X}_n - \bar{\mathbf{Y}}_0)$ and $\mathbf{Z}_n^{(2)} = h_n^{-1/2} \mathbf{P}_n \mathbf{C}_n^{-1} \mathbf{V}_n$, respectively. By using Lemma 3.1, we have $\mathbf{E}[\mathbf{Z}_n^{(1)} \mathbf{e}_g] = \mathbf{0}$, $\mathbf{E}[\mathbf{Z}_n^{(2)} \mathbf{e}_g] = \mathbf{0}$ and

$$\mathscr{E}[\mathbf{Z}_n^{(2)} \mathbf{e}_g \mathbf{e}_h' \mathbf{Z}_n^{(2)'}] = (\mathbf{e}_g' \Sigma_v \mathbf{e}_h) h_n^{-1} \mathbf{P}_n \mathbf{C}_n^{-1} \mathbf{C}_n'^{-1} \mathbf{P}_n = (\mathbf{e}_g' \Sigma_v \mathbf{e}_h) h_n^{-1} \mathbf{D}_n . \tag{5.17}$$

In the following derivations, we mainly discuss the estimation of integrated variance (or integrated volatility) because the estimation of integrated covariance is quite similar but with additional notation. One important difference is that we use the fact that in the limiting distribution $2(\mathbf{E}[X_g^2])^2$ should be replaced by $(\mathbf{E}[X_g^2])(\mathbf{E}[X_h^2]) + (\mathbf{E}[X_g X_h])^2$. (It is analogous to the standard practice in statistical multivariate analysis when $\mathbf{X} = (X_g)$ follows a multivariate normal distribution for any $g, h = 1, \ldots, p$. See Chap. 2 of Anderson 2003, for instance.) In our proofs of theorems we make an extensive use of the decomposition:

$$\hat{\sigma}_{gg}^{(x)} - \sigma_{gg}^{(x)} \tag{5.18}$$

$$= \frac{1}{m_n} \sum_{k=1}^{m_n} \left[z_{kg}^2 - \sigma_{gg}^{(x)} \right]$$

$$= \frac{1}{m_n} \sum_{k=1}^{m_n} \left[z_{kg}^{(1)2} - \sigma_{gg}^{(x)} + \sigma_{gg}^{(v)} a_{kn} \right] + \frac{1}{m_n} \sum_{k=1}^{m_n} \left[z_{kg}^{(2)2} - \sigma_{gg}^{(v)} a_{kn} \right]$$

$$+ 2 \frac{1}{m_n} \sum_{k=1}^{m_n} \left[z_{kg}^{(1)} z_{kg}^{(2)} \right] .$$

Lemma 5.3 *Assume the conditions of Theorem 3.3.*
(i) For $0 < \alpha < 0.5$,

$$\hat{\sigma}_{gg}^{(x)} - \sigma_{gg}^{(x)} \xrightarrow{p} 0 \tag{5.19}$$

as $n \to \infty$.

(ii) For $0 < \alpha < 0.4$,

$$\sqrt{m_n}\left[\hat{\sigma}_{gg}^{(x)} - \sigma_{gg}^{(x)} - \sum_{i=1}^{n}(c_{ii}-1)\int_{t_{i-1}}^{t_i}\sigma_x^2(s)ds\right] \tag{5.20}$$

$$-\sqrt{m_n}\left[\frac{1}{m}\sum_{k=1}^{m}\left(z_{kg}^{(1)2}\right) - \sigma_{gg}^{(x)} - \sum_{i=1}^{n}(c_{ii}-1)\int_{t_{i-1}}^{t_i}\sigma_x^2(s)ds\right]$$

$$\xrightarrow{p} 0$$

as $n \to \infty$.

Proof of Lemma 5.3: By using (3.15) and the relation $\sin x = x - (1/6)x^3 + (1/120)x^5 + O(x^7)$,

$$\frac{1}{m_n}\sum_{k=1}^{m_n} a_{kn} = \frac{1}{m_n}2n\sum_{k=1}^{m_n}\left[1 - \cos\left(\pi\frac{2k-1}{2n+1}\right)\right]$$

$$= \frac{n}{m_n}\left[2m_n - \frac{\sin\pi\frac{2m_n}{2n+1}}{\sin\pi\frac{1}{2n+1}}\right]$$

$$\sim \frac{n}{m_n}\left[2m_n - \frac{\left(\pi\frac{2m_n}{2n+1}\right) - \frac{1}{6}\left(\pi\frac{2m_n}{2n+1}\right)^3}{\left(\frac{\pi}{2n+1}\right) - \frac{1}{6}\left(\frac{\pi}{2n+1}\right)^3}\right]$$

$$= \frac{n}{m_n}\left[\frac{\frac{\pi^3}{6}\left(\frac{2m_n}{2n+1}\right)^3 - \frac{2\pi^2}{6}\left(\frac{1}{2n+1}\right)^3}{\frac{\pi}{2n+1}\left(1 - \frac{\pi^2}{6}\left(\frac{1}{2n+1}\right)^2\right)}\right],$$

which is approximately $(\pi^2/3)(m_n^2/n)$ and it is $O(\frac{m_n^2}{n})$. Also we find that

$$\frac{1}{m_n}\sum_{k=1}^{m_n} a_{kn}^2 = \frac{1}{m_n}4n^2\sum_{k=1}^{m_n}\left[1 - 2\cos\left(\pi\frac{2k-1}{2n+1}\right) + \frac{1}{2}\left(1 + \cos\left(2\pi\frac{2k-1}{2n+1}\right)\right)\right]$$

$$= \frac{4n^2}{m_n}\left[\frac{3}{2}m_n - \frac{\sin\pi\frac{2m_n}{2n+1}}{\sin\pi\frac{1}{2n+1}} + \frac{1}{4}\frac{\sin\pi\frac{4m_n}{2n+1}}{\sin\pi\frac{2}{2n+1}}\right],$$

which is $O(\frac{m_n^4}{n^2})$ as $n \to \infty$. Then the above terms are $o(1)$ when we have the condition that $m_n^2/n \to 0$ $(n \to \infty)$. Hence for the first term of (5.18) we need $0 < \alpha < 0.5$ for consistency and $0 < \alpha < 0.4$ for asymptotic normality as the minimum requirements because $(1/\sqrt{m_n})\sum_{i=1}^{m_n} a_{kn} = O(\sqrt{m_n}m_n^2/n)$ is negligible in the latter case. To show that these conditions are sufficient, we evaluate each term of $\sqrt{m_n}[\hat{\sigma}_{gg}^{(x)} - \sigma_{gg}^{(x)}]$ based on the decomposition (5.18).

For the third term of (5.18), there exists a positive constant K_1

$$\mathbf{E}\left[\frac{1}{\sqrt{m_n}}\sum_{k=1}^{m_n} z_{kg}^{(1)} z_{kg}^{(2)}\right]^2 = \frac{1}{m_n}\sum_{k,k'=1}^{m_n} \mathbf{E}\left[z_{kg}^{(1)} z_{k',g}^{(1)} z_{kg}^{(2)} z_{k',g}^{(2)}\right]$$

$$= \frac{1}{m_n}\sum_{k,k'=1}^{m_n} \mathbf{E}\left[2\sum_{j,j'=1}^{n} s_{jk} s_{j'k'} \mathbf{E}(r_{jg} r_{j',g}|\mathscr{F}_{n,\min(j,j')}) z_{kg}^{(2)} z_{k',g}^{(2)}\right]$$

$$= \frac{1}{m_n}\sum_{k,k'=1}^{m_n} \mathbf{E}\left[2\sum_{j=1}^{n} s_{jk} s_{j,k'} \mathbf{E}(r_{jg}^2|\mathscr{F}_{n,j-1}) z_{kg}^{(2)} z_{k',g}^{(2)}\right]$$

$$\leq K_1 \mathbf{E}\left[\sup_{0\leq s\leq 1} \sigma_{gg}^{(x)}(s)\right]^2 \frac{2}{n}\left(\frac{n}{2}+\frac{1}{4}\right)\frac{1}{m_n}\sum_{k=1}^{m_n} a_{kn},$$

which is $O(\frac{m_n^2}{n})$ and $\mathbf{\Sigma}_x(s) = (\sigma_{gh}^{(x)}(s))$. In the above evaluation we have used the independence of $x_{kg}^{(1)}$ and $x_{kg}^{(2)}$ and (5.9).

For the second term of (5.18), let $\mathbf{b}_k' = h_n^{-1/2}\mathbf{e}_k'\mathbf{P}_n'\mathbf{C}_n^{-1} = (b_{kj})$ and $\mathbf{e}_k' = (0,\ldots,1,0,\ldots)$ is an $n\times 1$ vector. Then we can write $z_{kg}^{(2)} = \sum_{j=1}^{n} b_{kj} v_{jg}$ and

$$\mathbf{E}\left[\frac{1}{\sqrt{m_m}}\sum_{k=1}^{m_n} (z_{kg}^{(2)2} - \sigma_{gg}^{(v)} a_{kn})\right]^2 \tag{5.21}$$

$$= \frac{1}{m_n}\sum_{k,k'=1}^{m_n} \mathbf{E}\left[(z_{kg}^{(2)2} - \sigma_{gg}^{(v)} a_{kn})(z_{k',g}^{(2)2} - \sigma_{gg}^{(v)} a_{k'n})\right]$$

$$= \frac{1}{m_n}\sum_{k,k'=1}^{m_n} \mathbf{E}\left[\left(\sum_{j=1}^{n} b_{kj} v_{jg}\right)^2 \left(\sum_{j'=1}^{n} b_{k'j'} v_{j',g}\right)^2 - \sigma_{gg}^{(v)} a_{kn} a_{k'n}\right]$$

$$\leq K_2 \frac{1}{m_n}\sum_{k=1}^{m_n}\sum_{j=1}^{n} b_{kj}^4,$$

which is less than $K_2(1/m_n)\sum_{k=1}^{m_n} a_{kn}^2 = O(m_n^4/n^2)$ for a positive constant K_2. Hence we have found that the main effect of the sampling errors associated with the SIML estimator of the integrated variance is the first term of (5.18).

Then we shall show the consistency of (5.19) and it means (3.22) with $g = h$. (For the covariances $\sigma_{gh}^{(x)}$ ($g, h = 1,\ldots,p$), we have the same derivation with more notations.)

We write $\mathbf{r}_i = (r_{ig}) = \mathbf{x}_i - \mathbf{x}_{i-1}$ ($i, j = 1,\ldots,n; g = 1,\ldots,p$) and by using the fact that $\mathbf{r}_i = (r_{ig})$ ($i = 1,\ldots,n; g = 1,\ldots,p$) are a sequence of martingale differences,

$$\mathbf{E}\left[\frac{1}{m_n}\sum_{k=1}^{m_n}(z_{kg}^{(1)2}-\sigma_{gg}^{(x)})\right]^2$$

$$=\left[\frac{2n}{2n+1}\right]^2\mathbf{E}\left\{\sum_{i,j=1}^{n}\left[c_{ij}\,r_{ig}r_{jg}-\delta_{ij}\int_{t_{i-1}}^{t_i}\sigma_{gg}^{(x)}(s)ds\right]\right\}^2+o(1)$$

$$=\left[\frac{2n}{2n+1}\right]^2\left(\mathbf{E}\left\{\sum_{i=j=1}^{n}\left[c_{ij}r_{ig}^2-\int_{t_{i-1}}^{t_i}\sigma_{gg}^{(x)}(s)ds\right]\right\}^2+\mathbf{E}\left\{\sum_{i\neq j=1}^{n}\left[c_{ij}r_{ig}r_{jg}\right]\right\}^2\right),$$

where $\delta_{ij}=1$ $(i=j)$; $\delta_{ij}=0$ $(i\neq j)$. Because $2n/(2n++1)\rightarrow 1$ as $n\rightarrow\infty$, we can ignore the factor $[2n/(2n+1)]^2$ and we need to evaluate

$$\mathbf{E}\left\{\sum_{i=1}^{n}\left[c_{ii}r_{ig}^2-\int_{t_{i-1}}^{t_i}\sigma_{gg}^{(x)}(s)ds\right]\right\}^2=\mathbf{E}\left[\sum_{i=1}^{n}(r_{ig}^2-\int_{t_{i-1}}^{t_i}\sigma_{gg}^{(x)}(s)ds)+\sum_{i=1}^{n}(c_{ii}-1)r_{ig}^2\right]^2,$$

$$\mathbf{E}\left\{\left[\sum_{i\neq j=1}^{n}c_{ij}r_{ig}r_{jg}\right]^2\right\}=2\sum_{i\neq j=1}^{n}c_{ij}^2\mathscr{E}(r_{ig}^2)\mathscr{E}(r_{jg}^2)\,. \tag{5.22}$$

We also have the relation that for a constant K_3,

$$\mathbf{E}(r_{ig}^2)=\mathbf{E}(\int_{t_{i-1}}^{t_i}[c_{gg}^{(x)}(s)]^2ds)\leq\frac{K_3}{n}\,.$$

Hence the first term of (5.22) is approximately equivalent to

$$\sum_{i=1}^{n}\left[r_{ig}^2-\int_{t_{i-1}}^{t_i}\sigma_{gg}^{(x)}(s)ds+(c_{ii}-1)r_{ig}^2\right]$$

$$=\sqrt{\frac{1}{n}}\sqrt{n}\sum_{i=1}^{n}\left[r_{ig}^2-\int_{t_{i-1}}^{t_i}\sigma_{gg}^{(x)}(s)ds\right]+\left[\sum_{i=1}^{n}(c_{ii}-1)(r_{ig}^2-\int_{t_{i-1}}^{t_i}\sigma_{gg}^{(x)}(s)ds)\right]$$

$$+\left[\sum_{i=1}^{n}(c_{ii}-1)\int_{t_{i-1}}^{t_i}\sigma_{gg}^{(x)}ds\right]\,.$$

We have the basic relation (Jacod and Protter 1998 for instance)

$$\sqrt{n}\sum_{i=1}^{n}\left[r_{ig}^2-\int_{t_{i-1}}^{t_i}\sigma_x^2(s)ds\right]=O_p(1)\,. \tag{5.23}$$

and the inequality

$$|\sum_{i=1}^{n}(c_{ii}-1)\int_{t_{i-1}}^{t_i}\sigma_{gg}^{(x)}(s)ds|^2 = \left[\sum_{i=1}^{n}(c_{ii}-1)^2\right]\left[\sum_{i=1}^{n}\left(\int_{t_{i-1}}^{t_i}\sigma_{gg}^{(x)}(s)ds\right)^2\right]$$

$$\leq \left[\frac{1}{n}\sum_{i=1}^{n}(c_{ii}-1)^2\right]\left[\sup_{0\leq s\leq t}\sigma_{gg}^{(x)}(s)\right]^2$$

$$= O_p\left(\frac{1}{m_n}\right)$$

and

$$\mathbf{E}\left[|\sum_{i=1}^{n}(c_{ii}-1)(r_{ig}^2 - \int_{t_{i-1}}^{t_i}\sigma_{gg}^{(x)}(s)ds)|^2\right] = \left[\sum_{i=1}^{n}(c_{ii}-1)^2\mathbf{E}(r_{ig}^2 - \int_{t_{i-1}}^{t_i}\sigma_{gg}^{(x)}(s)ds)^2\right]$$

$$= O\left(\frac{1}{n}\right).$$

Hence by using (5.11) of Lemma 5.2 to the first term of (5.22), we have (5.19) and (5.20) when $m_n/n \to 0$ as $n \to \infty$. $\qquad\square$

Lemma 5.4 *For $0 < \alpha \leq 0.4$,*

$$\sqrt{m_n}\left[\sum_{i=1}^{n}(c_{ii}-1)\int_{t_{i-1}}^{t_i}\sigma_{gg}^{(x)}(s)ds\right] \xrightarrow{p} 0 \qquad (5.24)$$

as $n \to \infty$.

Proof of Lemma 5.4: We use the relation

$$\sqrt{m_n}\left[\sum_{i=1}^{n}(c_{ii}-1)\int_{t_{i-1}}^{t_i}\sigma_{gg}^{(x)}(s)ds\right]$$

$$= \sum_{i=1}^{n}\frac{1}{2\sqrt{m_n}}\left[\frac{\sin\left[\frac{2\pi m_n}{2n+1}(2i-1)\right]}{\sin\left[\frac{\pi}{2n+1}(2i-1)\right]}\right]\int_{t_{i-1}}^{t_i}\sigma_{gg}^{(x)}(s)ds .$$

We take a positive constant γ $(0 < \gamma < 1)$ and divide the summation of the right-hand side from 1 to n into two parts, that is, (i) $1 \leq i \leq n^\gamma$ and (ii) $n^\gamma + 1 \leq i \leq n$. For (i) there exists a positive K_4 such that the first part of the summation is less than

$$K_4\frac{1}{\sqrt{m_n}}\sum_{i=1}^{n^\gamma}\frac{n}{i}\left[\frac{1}{n}\sup_{0\leq s\leq 1}\sigma_{gg}^{(x)}(s)\right] = O\left(\frac{\log n^\gamma}{\sqrt{m_n}}\right) . \qquad (5.25)$$

For (ii) there exists a positive K_5 such than the second part of the summation is less than

$$K_5\frac{1}{\sqrt{m_n}}\sum_{i=n^\gamma+1}^{n}\frac{n}{n^\gamma}\left[\frac{1}{n}\sup_{0\leq s\leq 1}\sigma_{gg}^{(x)}(s)\right] = O\left(\frac{n}{n^{\gamma+\alpha/2}}\right) . \qquad (5.26)$$

Hence if we impose the condition $\gamma + \alpha/2 > 1$, both terms converge to zero as $n \longrightarrow \infty$. (We can take γ satisfying this condition.) $\qquad\qquad\square$

Lemma 5.5 *Under the assumptions of Theorem 3.1 (or Theorem 3.3) with the condition $0 < \beta < 1$, as $n \to \infty$,*

$$\hat{\sigma}_{gh}^{(v)} - \sigma_{gh}^{(v)} \xrightarrow{p} 0 \tag{5.27}$$

and

$$\sqrt{l_n} \left[\hat{\sigma}_{gh}^{(v)} - \sigma_{gh}^{(v)} \right] = O_p(1) . \tag{5.28}$$

Proof of Lemma 5.5: We give only a brief proof of how to estimate the noise variance $\sigma_{gg}^{(v)}$ because it is quite similar to the estimation of the noise covariance. For this purpose we use the decomposition

$$\hat{\sigma}_{gg}^{(v)} - \sigma_{gg}^{(v)} \tag{5.29}$$

$$= \frac{1}{l_n} \sum_{k=n+1-l}^{n} a_{kn}^{-1} \left[z_{kg}^2 - \sigma_{gg}^{(x)} a_{kn} \right]$$

$$= \frac{1}{l_n} \sum_{k=n+1-l_n}^{n} a_{kn}^{-1} \left[z_{kg}^{(2)2} - \sigma_{gg}^{(v)} a_{kn} \right] + \frac{1}{l_n} \sum_{k=n+1-l_n}^{n} a_{kn}^{-1} \left[z_{kg}^{(1)2} + 2 z_{kg}^{(1)} z_{kg}^{(2)} \right] .$$

Then the main argument of the proof is similar to that of Lemma 5.3 except with l_n instead of m_n. For the variance of the noise term, we use the fact that $l_n/n = o(1)$ and for $n + 1 - l_n \le k \le n$ and $a_{kn} = 2n[1 + \cos \pi(\frac{2l_n}{2n+1})] \ge n$ for a sufficiently large n. Because $a_{kn}^{-1} = o(n^{-1})$,

$$\mathbf{E}[\sum_{k=n+1-l_n}^{n} a_{kn}^{-1}(z_{kg}^{(1)2})] = \sigma_{gg}^{(x)} \sum_{k=n+1-l_n}^{n} a_{kn}^{-1} = O\left(\frac{l_n}{n}\right) . \tag{5.30}$$

Then by using the similar evaluations as

$$\mathbf{E}\left[\frac{1}{\sqrt{l_n}} \sum_{k=n+1-l_n}^{n} a_{kn}^{-1} z_{kg}^{(1)2} \right]^2 = o(1)$$

and

$$\mathbf{E}\left[\frac{1}{\sqrt{l_n}} \sum_{k=n+1-l_n}^{n} a_{kn}^{-1} z_{kg}^{(1)} z_{kg}^{(2)} \right]^2 = o(1) . \tag{5.31}$$

Hence we can ignore the last two terms on the right-hand side of (5.29) and we need only evaluate the leading term. Then by using a similar evaluation as (5.18), it is possible to evaluate

$$\mathbf{E}\left[\frac{1}{l_n}\sum_{k=n+1-l_n}^{n}a_{kn}^{-1}\left(z_{kg}^{(2)2}-\sigma_{gg}^{(v)}a_{kn}\right)\right]^2 = o(1) \tag{5.32}$$

and

$$\mathbf{E}\left[\frac{1}{\sqrt{l_n}}\sum_{k=n+1-l_n}^{n}a_{kn}^{-1}\left(\sum_{i,j=1;i\neq j}^{n}b_{ik}b_{jk}v_{ig}v_{jg}\right)\right]^2 = O(1). \tag{5.33}$$

$\square$

Proof of Theorem 3.3: (**Step 1**) the of the SIML estimator, we use Lemmas 5.4 and 5.5 to have the result.

Then we prove the asymptotic normality of the SIML estimator of integrated volatility $\sigma_{gh}^{(x)}$ $(g, h = 1, \ldots, p)$. First, we need to derive the asymptotic variance of the SIML estimator. We write $\sqrt{m_n}[\hat{\sigma}_{gh}^{(x)} - \sigma_{gh}^{(x)}]$ as

$$\sqrt{m}\left[\sum_{i,j=1}^{n}\left(c_{ij}r_{ig}r_{jh}-\delta_{ij}\int_{t_{i-1}}^{t_i}\sigma_{gh}^{(x)}(s)ds\right)\right] \tag{5.34}$$

$$= \sqrt{m}\sum_{i>j}c_{ij}[r_{ig}r_{jh}+r_{jg}r_{ih}]+\sqrt{m}\left[\sum_{i=1}^{n}c_{ii}r_{ig}r_{ih}-\int_{t_{i-1}}^{t_i}\sigma_{gh}^{(x)}(s)ds\right],$$

where $\delta_{ij} = 1$ $(i = j)$; $\delta_{ij} = 0$ $(i \neq j)$.

First, we have the relation

$$\mathbf{E}[\sqrt{m_n}\sum_{i>j}c_{ij}(r_{ig}r_{jh}+r_{jg}r_{ih})]^2 = m_n\sum_{i>j}c_{ij}^2\mathbf{E}\left[r_{ig}^2r_{jh}^2+r_{jg}^2r_{ih}^2+2r_{ig}r_{jh}r_{jg}r_{ih}\right]$$

$$= m_n\sum_{i\neq j}c_{ij}^2\left[\mathbf{E}(r_{ig}^2r_{jh}^2)+\mathbf{E}(r_{ig}r_{ih}r_{jg}r_{jh})\right].$$

By using the notation $\mathbf{E}_{i-1}[r_{ig}^2] = \mathbf{E}[r_{ig}^2|\mathscr{F}_{n,i-1}]$,

$$\sum_{i=1}^{n}m_nc_{ii}^2\left(\mathbf{E}_{i-1}[r_{ig}^2]\right)^2 \leq \left[\mathbf{E}\sup_{0\leq s\leq 1}\sigma_{gg}^{(x)}(s)\right]^2\frac{m_n}{n^2}\sum_{i=1}^{n}c_{ii}^2 \to 0 \tag{5.35}$$

as $m_n/n \to 0$. Then, the asymptotic variance is the limit of

$$V_{gh.n} = \sum_{i,j=1}^{n}m_nc_{ij}^2\left[\int_{t_{i-1}}^{t_i}\tilde{\sigma}_{gg}^{(x)}(s)ds\int_{t_{j-1}}^{t_j}\sigma_{hh}^{(x)}(s)ds+\int_{t_{i-1}}^{t_i}\sigma_{gh}^{(x)}(s)ds\int_{t_{j-1}}^{t_j}\sigma_{gh}^{(x)}(s)ds\right].$$

$$\tag{5.36}$$

Lemma 5.6 *As* $n \to \infty$

$$V_{gh.n} \xrightarrow{P} V_{gh} = \int_0^1 \left[\sigma_{gg}^{(x)}(s)\sigma_{hh}^{(x)}(s) + \sigma_{gh}^{(x)2}(s) \right] ds \,. \tag{5.37}$$

Proof of Lemma 5.6: We set $m = m_n$ and by using Lemma 5.1, we find the relation

$$m \, c_{ij}^2 = \frac{1}{4m} \left[\frac{\sin \pi m \left(\frac{i+j-1}{n+1/2} \right)}{\sin(\pi/2) \left(\frac{i+j-1}{n+1/2} \right)} + \frac{\sin \pi m \left(\frac{i-j1}{n+1/2} \right)}{\sin(\pi/2) \left(\frac{i-j}{n+1/2} \right)} \right]^2 \,. \tag{5.38}$$

For any real λ and ν, let

$$k_m(\lambda, \nu) = \frac{1}{2\pi m} \frac{\sin \frac{\lambda m}{2} \sin \frac{\nu m}{2}}{\sin \frac{\lambda}{2} \sin \frac{\nu}{2}}, \quad k_m(\lambda) = \frac{1}{2\pi m} \frac{\sin^2 \frac{\lambda m}{2}}{\sin^2 \frac{\lambda}{2}},$$

and we use the classical evaluation method of the Fejé-kernel due to Chap. 8 of Anderson (1971). The effects of discretization of continuous functions are negligible in the present case. Then for the square-integrable continuous functions $a(s)$ and $b(s)$, we write

$$\frac{1}{4m} \int_0^1 \int_0^1 \left[\frac{\sin \pi m(s+t)}{\sin \frac{\pi}{2}(s+t)} + \frac{\sin \pi m(s-t)}{\sin \frac{\pi}{2}(s-t)} \right]^2 a(s)b(t)ds dt \tag{5.39}$$

$$= \frac{1}{4m} \int_0^1 \int_0^1 \left(\left[\frac{\sin \pi m(s+t)}{\sin \frac{\pi}{2}(s+t)} \right]^2 + \left[\frac{\sin \pi m(s-t)}{\sin \frac{\pi}{2}(s-t)} \right]^2 \right.$$

$$\left. +2 \left[\frac{\sin \pi m(s+t)}{\sin \frac{\pi}{2}(s+t)} \right] \left[\frac{\sin \pi m(s-t)}{\sin \frac{\pi}{2}(s-t)} \right] \right) a(s)b(t)ds dt \,.$$

By using a transformation the second term becomes

$$\int_0^1 \frac{\sin^2 \frac{\pi}{2} 2mu}{2m \sin^2 \frac{\pi}{2} u} \left[\int_0^{1-u} a(u+t)b(t)dt \right] du \,. \tag{5.40}$$

Then by using the standard evaluation method used in Chap. 8 of Anderson (1971) with $k_m(\lambda)$, we find that the second term converges to $\int_0^1 a(t)b(t)dt$ as $m \to \infty$ while the other two terms converge to zeros. Finally by applying the above arguments to each term in the present case we have the result. We have omitted some details because they may be straightforward.   □

(**Step 2**) Next, we need to show that the normalized SIML estimator converges to the limiting Gaussian random variable when $g = h$.

In our proof we use of a sequence of random variables

$$U_n = \sum_{j=2}^{n} \left[2 \sum_{i=1}^{j-1} \sqrt{m_n} c_{ij} r_{ig} \right] r_{jg}, \tag{5.41}$$

which is a discrete martingale. We notice that for the process $\mathbf{x}(t)$ in (3.2) and $\mathbf{r}_i = (r_{ig}) = \mathbf{x}_i - \mathbf{x}_{i-1}$ $(i = 1, \ldots, n)$ are the (discrete) martingale parts. Then we set $X_{nj} = (2 \sum_{i=1}^{j-1} \sqrt{m_n} c_{ij} r_{ig}) r_{jg}$, $W_{nj} = r_{jg}$ $(j = 2, \ldots, n)$, $V_{gg.n}^* = \sum_{j=2}^{n} \mathbf{E}[X_{nj}^2 | \mathcal{F}_{n,j-1}]$, and we apply the martingale central limit theorem (MCLT). Thus we have to check Condition (B)

$$\sum_{j=1}^{n} \mathbf{E}[X_{nj}^4] \longrightarrow 0 , \tag{5.42}$$

and Condition (C)

$$\mathbf{E}[(V_{gg.n}^* - V_{gg})^2] \longrightarrow 0 \tag{5.43}$$

as $n \longrightarrow \infty$. It it because we have Lemma 5.6 and Condition (B) implies the Lindeberg Condition such that $\sum_{j=1}^{n} \mathbf{E}[X_{nj}^2 I(|X_{nj}| > \epsilon)] \longrightarrow 0$ as $n \to \infty$ for any ϵ (> 0). Then it is possible to use Theorem 35.12 of Billingsley (1995) as the MCLT.

In the above conditions, first Lemma 5.7 below shows Condition (B).

Second, under the assumptions we have $V_{gg.n} \xrightarrow{p} V_{gg}$ where $V_{gg.n}$ is given in Lemma 5.6. We note that in the present situation that $V_{gg.n}$ and V_{gg} are bounded. Then we can find a positive K_6 such that for any $\epsilon > 0$

$$\begin{aligned} \mathbf{E}[(V_{gg.n} - V_{gg})^2] &= \mathbf{E}[(V_{gg.n} - V_{gg})^2 I(|V_{gg.n} - V_{gg}| \geq \epsilon)] \\ &\quad + \mathbf{E}[(V_{gg.n} - V_{gg})^2 I(|V_{gg.n} - V_{gg}| < \epsilon)] \\ &\leq K_6 \mathbf{P}(|V_{gg.n} - V_{gg}| \geq \epsilon) + \epsilon^2 . \end{aligned}$$

Hence we need only to show Condition (D)

$$\mathbf{E}[(V_{gg.n}^* - V_{gg.n})^2] \longrightarrow 0 \tag{5.44}$$

as $n \longrightarrow \infty$. Then Lemma 5.8 below shows Condition (D).

(**Step 3**) For the case of $\hat{\sigma}_{gh}$ and $V_{gh.n}$ with $g \neq h$, we use the similar arguments but with some more notations. Also in order to show the joint normality of the limiting random variables, we can use the standard device in the statistical multivariate analysis. We have omitted the details of these routine works.   □

Lemma 5.7 *Under a set of assumptions, we have Condition (B).*

Proof of Lemma 5.7: Without loss of generality we consider the stochastic case when $p = q = 1$ and we denote $\mathbf{C}_x(s) = c_s$. When the instantaneous volatility function is time-varying and bounded function as the assumption of Theorem 3.3, the proof is the result of straightforward calculation.

Here we shall show Condition (B) under a set of assumptions, which includes stochastic cases. We set

$$Z_{nj}(t) = \int_{t_{j-1}}^{t} c_s dB_s \ (t_{j-1} \le t \le t_j, j = 1, \ldots, n)$$

and

$$W_{nj}^{*} = \sum_{i=1}^{j-1} \sqrt{m_n} c_{ij} \int_{t_{i-1}}^{t_i} c_s dB_s \ (j = 2, \ldots, n) \ .$$

Then we need to show that

$$\sum_{j=2}^{n} \mathbf{E}[W_{nj}^{*4} Z_{nj}(t_j)^4] \longrightarrow 0 \tag{5.45}$$

as $n \to \infty$.

First by using Ito's Lemma, we have

$$Z_{nj}(t)^4 = \int_{t_{j-1}}^{t} 4[Z_{nj}(s)]^3 c_s dB_s + \int_{t_{j-1}}^{t} 6[Z_{nj}(s)]^2 c_s^2 ds \ .$$

Then by taking the conditional expectation of both sides given $\mathscr{F}_{n,j-1}$ (we denote the conditional expectation as $\mathbf{E}[\cdot | \mathscr{F}_{n,j-1}] = \mathbf{E}_{j-1}[\cdot]$), we have

$$\mathbf{E}_{j-1}[Z_{nj}(t)^4] = \int_{t_{j-1}}^{t} 6\mathbf{E}_{j-1}[(Z_{nj}(t))^2 c_s^2] ds \ . \tag{5.46}$$

Also by using the inequality $\mathbf{E}[X^2 Y^2] \le (1/2)[\mathbf{E}(X^4) + \mathbf{E}(Y^4)]$

$$\mathbf{E}_{j-1}[Z_{nj}(t)^4] \le 3 \int_{t_{j-1}}^{t} \mathbf{E}_{j-1}[(Z_{nj}(s))^4] ds + 3 \int_{t_{j-1}}^{t} \mathbf{E}_{j-1}[c_s^4] ds \ . \tag{5.47}$$

Furthermore, by using the boundedness condition, we have $\int_{t_{j-1}}^{t} c_s^4 ds = O(\frac{1}{n})$. Then by using the standard argument in stochastic calculus regarding the evaluation of moments (i.e., Chap. 3 of Ikeda and Watanabe (1989), for instance), we can find a positive constant K_7 such that $\mathbf{E}_{j-1}[Z_{nj}(t)^4] \le K_7(1/n)$. By applying the Cauchy-Schwartz inequality to (5.48), we can find a positive constant K_8 such that

$$\mathbf{E}_{j-1}[Z_{nj}(t)^4] \le 6 \int_{t_{j-1}}^{t} \left(\mathbf{E}_{j-1}[Z_{nj}^4(s)]\right)^{1/2} \left(\mathbf{E}_{j-1}[c_s^4]\right)^{1/2} ds \le K_8 \left[\frac{1}{n}\right]^{1+\frac{1}{2}} \ .$$

By repeating the above substitution procedure, we have the bound of the fourth-order moment as $K_8'(\frac{1}{n})^{1+1/2+(1/2)^2+\cdots+(1/2)^r}$ for an arbitrary positive integer r $(r \ge 2)$ and a positive constant K_8'. Then we can find that for an arbitrary small ϵ (> 0) and $t_{j-1} \le t \le t_j$, $\mathbf{E}_{j-1}[Z_{nj}(t)^4] = O((1/n^{2(1-\epsilon)})$.

Next, we evaluate the expectation $\mathbf{E}[W_{nj}(t)^{*4}]$, that is

$$
\begin{aligned}
&\mathbf{E}[W_{nj}(t)^{*4}] \\
&= \mathbf{E}\left[\sum_{i_1,i_2,i_3,i_4=1}^{j-1} m_n^2 c_{i_1,j} c_{i_2,j} c_{i_3,j} c_{i_4,j} \int_{t_{i_1-1}}^{t_{i_1}} c_{s_1} dB_{s_1} \int_{t_{i_2-1}}^{t_{i_2}} c_{s_2} dB_{s_2} \right. \\
&\qquad \left. \int_{t_{i_3-1}}^{t_{i_3}} c_{s_3} dB_{s_3} \int_{t_{i_4-1}}^{t_{i_4}} c_{s_4} dB_{s_4} \right].
\end{aligned}
$$

In this form we need only consider the summations of the forms (i) $\sum_{i_1=i_2,i_3,i_4}[\,\cdot\,]$ and (ii) $\sum_{i_1=i_2,i_3=i_4}[\,\cdot\,]$ because $\int_{t_{i_1-1}}^{t_{i_1}} c_{s_1} dB_{s_1}$ is a martingale difference. We first consider Case (i) and we set $i_1 = i_2 > i_3 > i_4$. In this case we can utilize the fact that $\mathbf{E}[\int_{t_{i_1-1}}^{t_{i_1}} c_{s_1} dB_{s_1}]^2 = \int_{t_{i_1-1}}^{t_{i_1}} c_{s_1}^2 ds$ and $|2\int_{t_{i_3-1}}^{t_{i_3}} c_{s_3} dB_{s_3} \int_{t_{i_4-1}}^{t_{i_4}} c_{s_4} dB_{s_4}| \le [\int_{t_{i_3-1}}^{t_{i_3}} c_{s_3} dB_{s_3}]^2 + [\int_{t_{i_4-1}}^{t_{i_4}} c_{s_4} dB_{s_4}]^2$.

By using the assumption that c_s are bounded and $\int_{t_{i_1-1}}^{t_{i_1}} c_{s_1} dB_{s_1}$ is a martingale difference, we can find a positive constant K_9 such that

$$
\mathbf{E}\left[|\int_{t_{i_1-1}}^{t_{i_1}} c_{s_1} dB_{s_1} \int_{t_{i_2-1}}^{t_{i_2}} c_{s_2} dB_{s_2} \int_{t_{i_3-1}}^{t_{i_3}} c_{s_3} dB_{s_3} \int_{t_{i_4-1}}^{t_{i_4}} c_{s_4} dB_{s_4}| \right] \le K_9 \left(\frac{1}{n}\right)^2 . \quad (5.48)
$$

Hence we have $\mathbf{E}[W_{nj}(t)^{*4}] = o(1/n)$ if we can show

$$
\left[\sum_{i_1,i_3,i_4}^{j-1} m_n^2 c_{i_1,j}^2 c_{i_3,j} c_{i_4,j} \left(\frac{1}{n}\right)^{1+2(1-\epsilon)} \right] \longrightarrow 0 \quad (5.49)
$$

as $n \to \infty$. We apply the method used to prove Lemma 5.4 and for $l > 1$ we use the relation

$$
\sum_{i=1}^{l} c_{ij} = \frac{2}{m_n} \sum_{k=1}^{m_n} \left[\frac{\sin\left[\frac{\pi l}{2n+1}(2k-1)\right]}{2\sin\left[\frac{\pi}{2n+1}(2k-1)\right]} \right] s_{jk} .
$$

For $m_n = n^\alpha$ and $k_n = n^\gamma$ ($\alpha + \gamma - 1 > 0$) we have $\sin\left[\frac{\pi l}{2n+1}(2k_n-1)\right]/\sin\left[\frac{\pi l}{2n+1}(2k_n-1)\right] \to 0$ ($n \to \infty$) and we use similar arguments as those in Lemma 5.4. Then for a sufficiently small ϵ (> 0) we have

$$
\frac{\sqrt{m_n}}{n^{1-\epsilon}} \sum_{i=1}^{l} c_{ij} = o(1) \quad (5.50)
$$

as $n \to \infty$. Hence we have the result for Case (i). We can use the same evaluation method for Case (ii), whereupon we obtain the order of $\mathbf{E}[W_{nj}(t)^{*4}]$.

Finally, because we make the assumption that c_s and $\sigma_{gg}^{(x)}(s)$ are bounded and $W_{nj}^* Z_{nj}(t)$ $(t_{j-1} \leq t \leq t_j; j = 2, \dots, n)$ is a sequence of martingale differences, we evaluate the fourth-order moments similarly. Then we have the desired result. $\qquad\square$

Lemma 5.8 *Under a set of assumptions, we have Condition (D).*

Proof of Lemma 5.8: Without loss of generality, we consider the case in which $p = q = 1$. Here we shall show Condition (D) under a set of assumptions, which includes stochastic cases.

Then it is sufficient to evaluate

$$
D_n = \mathbf{E}\left\{\left(\sum_{j=1}^{n}[\mathbf{E}_{j-1}(X_{nj}^2) - \mathbf{E}(X_{nj}^2)]\right)^2\right\}
$$

$$
= \mathbf{E}\left\{\sum_{j=1}^{n}\left(\sum_{i_1,i_2=1}^{n} m_n c_{i_1,j} c_{i_2,j}\left(\int_{t_{j-1}}^{t_j} c_s^2 ds\right)\right.\right.
$$

$$
\left.\left.\left[\left(\int_{t_{i_1-1}}^{t_{i_1}} c_{s_{i_1}} dB_{s_{i_1}}\right)\left(\int_{t_{i_2-1}}^{t_{i_2}} c_{s_{i_2}} dB_{s_{i_2}}\right) - \delta(i_1,i_2)\left(\int_{t_{i_1-1}}^{t_{i_1}} c_{s_{i_1}}^2 ds_{i_1}\right)\right]\right)^2\right\},
$$

where $\mathbf{E}_{j-1}(X_{nj}^2) = \mathbf{E}(X_{nj}^2|\mathscr{F}_{n,j-1})$, and $\delta(i_1,i_2) = 1$ for $i_1 = i_2$ and $\delta(i_1,i_2) = 0$ for $i_1 \neq i_2$.

We use the arguments as (5.49) with $t_{i_1} = t_{i_2}, t_{i_1} = t_{i_3}$ or $t_{i_1} = t_{i_4}$ in the proof of Lemma 5.7. Because we have $\mathbf{E}[\int_{t_{i-1}}^{t_i} c_s dB_s]^2 = O_p(\frac{1}{n})$, which is bounded under the present formulation, and Lemma 5.2, there exist positive constants K_{10} and K_{11} such that

$$
D_n \leq K_{10}(\frac{1}{n})^2 \sum_{i_1,i_2=1}^{n}\left[\sum_{j,j'=1}^{n} m_n^2 c_{i_1,j} c_{i_2,j} c_{i_1,j'} c_{i_2,j'}\left(\int_{t_{j-1}}^{t_j} c_s^2 ds\right)\left(\int_{t_{j'-1}}^{t_{j'}} c_s^2 ds'\right)\right]
$$

$$
= K_{10}\left(\frac{1}{n}\right)^2\left[\sum_{j,j'=1}^{n}\left(\sum_{i_1=1}^{n} m_n c_{i_1,j} c_{i_1,j'}\right)\left(\sum_{i_2=1}^{n} m_n c_{i_2,j} c_{i_2,j'}\right)\left(\int_{t_{j-1}}^{t_j} c_s^2 ds\right)\left(\int_{t_{j'-1}}^{t_{j'}} c_s^2 ds'\right)\right]
$$

$$
\leq K_{11}\left(\frac{1}{n}\right)^2\left(n+\frac{1}{2}\right)^2\left[\sum_{j,j'=1}^{n} c_{j,j'}^2\right]\left(\frac{1}{n}\right)^2.
$$

Then by using Lemma 5.2 again and the fact that $\mathbf{E}[\int_{t_{j-1}}^{t_j} c_s dB_s]^2 = O(\frac{1}{n})$, finally we find that $D_n = O(\frac{1}{m_n})$. $\qquad\square$

Proof of Theorem 3.1: We give some additional arguments to the proofs of Theorem 3.3 and Lemma 5.5. By using $\mathbf{b}_k' = (b_{kj})$, $\sigma_{gh}^{(v)}$ and $\hat{\sigma}_{gh}^{(v)}$ for $g, h = 1, \dots, p$, we use the representation

$$\sqrt{l_n} \left[\frac{1}{l_n} \sum_{k=n+1-l_n}^{n} a_{kn}^{-1} \left(\sum_{i=1}^{n} b_{ki} v_{ig} \right) \left(\sum_{j=1}^{n} b_{kj} v_{jh} \right) - \sigma_{gh}^{(v)} \right] \tag{5.51}$$

$$= \frac{1}{\sqrt{l_n}} \sum_{k=n+1-l_n}^{n} a_{kn}^{-1} \left[\sum_{i=j=1}^{n} b_{ki}^2 (v_{ig} v_{ih} - \sigma_{gh}^{(v)}) \right] + \frac{1}{\sqrt{l_n}} \sum_{k=n+1-l_n}^{n} a_{kn}^{-1} \left[\sum_{i \neq j}^{n} b_{ki} b_{kj} v_{ig} v_{jh} \right].$$

We denote (I) and (II) for $\sqrt{l_n}$ times each terms of the right-hand side and evaluate their variances. The variance of (I) is given by

$$\mathbf{Var}(\mathrm{I}) = \sum_{i=1}^{n} \left[\sum_{k=n+1-l_n}^{n} a_{kn}^{-1} b_{ki} \right]^2 \mathbf{Var}(v_{ig} v_{ih})].$$

When $\mathbf{v}_i$ follows the Gaussian distribution, the last term becomes $\sigma_{gg}^{(v)} \sigma_{hh}^{(v)} + \sigma_{gh}^{(v)2}$ and then we can denote κ_{gh} as the effect of non-Gaussianity. (It turns out that the effects of κ_{gh} are asymptotically negligible.)

The variance of (II) can be written

$$\mathbf{Var}(\mathrm{II}) = \mathbf{E} \left[\sum_{i>j} \sum_{k} a_{kn}^{-1} b_{ki} b_{kj} v_{ig} v_{jh} + \sum_{i<j} \sum_{k} a_{kn}^{-1} b_{ki} b_{kj} v_{ig} v_{jh} \right]^2$$

$$= [\sigma_{gg}^{(v)} \sigma_{hh}^{(v)} + \sigma_{gh}^{(v)2}] \left[\sum_{i>j}^{n} \left(\sum_{k} a_{kn}^{-1} b_{ki} b_{kj} \right)^2 + \sum_{i<j}^{n} \left(\sum_{k} a_{kn}^{-1} b_{ki} b_{kj} \right)^2 \right].$$

The covariance of (I) and (II) is asymptotically negligible. Then, by using the fact that $\sum_{i=1}^{n} b_{ki} b_{k',i} = \delta(k, k') a_{kn} + O(1)$ and $\sum_{k=n+1-l_n}^{n} a_{kn} b_{ki}^2 = o(1)$ for $k = n + 1 - l_n, \ldots, n$, the variance of (5.51) is approximately given by $l_n[\sigma_{gg}^{(v)} \sigma_{hh}^{(v)} + \sigma_{gh}^{(v)2}]$. Also by applying MCLT, we have the asymptotic normality in (iv) of Theorem 3.1.

For the asymptotic covariances of random variables $\hat{\sigma}_{gh}^{(v)}$ and $\hat{\sigma}_{kl}^{(v)}$ ($g, h = 1, \ldots, p$), we have similar evaluations.

For the last part of (ii), we can evaluate the asymptotic covariances of random variables $\hat{\sigma}_{gh}^{(x)}$ and $\hat{\sigma}_{kl}^{(x)}$ ($g, h = 1, \ldots, p$), which are straightforward, and we omit the details. $\square$

Proof of Corollary 3.1: When $\sigma_{gg}^{(v)} = 0$, we have $\mathbf{X}_n^{(2)} \mathbf{e}_g = \mathbf{0}$ and then $\mathbf{Z}_n \mathbf{e}_g = h_n^{-1/2} \mathbf{P}_n \mathbf{C}_n (\mathbf{X}_n - \bar{\mathbf{Y}}_0) \mathbf{e}_g$. We use the relation

$$T_1 = \sqrt{m_n} \left[\frac{\left(\frac{1}{l_n} \sum_{k=n+1-l_n}^{n} z_{kg}^2 - \sigma_{gg}^{(x)} \right) - \left(\frac{1}{m_n} \sum_{k=1}^{m_n} z_{kg}^2 - \sigma_{gg}^{(x)} \right)}{\sigma_{gg}^{(x)} + \left(\frac{1}{m_n} \sum_{k=1}^{m_n} z_{kg}^2 - \sigma_{gg}^{(x)} \right)} \right]$$

$$= -\sqrt{m_n} \left[\frac{\frac{1}{m_n} \sum_{k=1}^{m_n} z_{kg}^2 - \sigma_{gg}^{(x)}}{\sigma_{gg}^{(x)}} \right] + \frac{\sqrt{m_n}}{\sqrt{l_n}} \sqrt{l_n} \left[\frac{\frac{1}{l_n} \sum_{k=n+1-l_n}^{n} z_{kg}^2 - \sigma_{gg}^{(x)}}{\sigma_{gg}^{(x)}} \right]$$

$$+ o_p(1).$$

Because of the condition $0 < \alpha < \beta < 1$, we have $m_n/l_n \to 0$ as $n \to \infty$ and then the second term of T_1 converges to zero in probability. The first term converges to $N(0, 2)$ by Theorem 3.1, and thus we have the result. $\qquad\square$

Finally, we give the proof of Theorem 3.4. Although we need to use the stable convergence, we omit its details because there are standard literatures available as we have stated.

Proof of Theorem 3.4: In addition to proving Theorem 3.3, we give some additional arguments that (i) the effect of the existence of drift term in the stochastic process is asymptotically negligible, and (ii) we have the same form of limiting random variables for the case of stochastic volatility.

(On Drift Terms) We write the returns as

$$\mathbf{r}_i^n = \mathbf{x}_i^n - \mathbf{x}_{i-1}^n = \int_{t_{i-1}^n}^{t_i^n} \boldsymbol{\mu}_x(s)ds + \int_{t_{i-1}^n}^{t_i^n} \mathbf{C}_x(s)d\mathbf{B}_s \quad (i = 1, \dots, n) \qquad (5.52)$$

and the martingale part as

$$\mathbf{r}_i^* = \int_{t_{i-1}^n}^{t_i^n} \mathbf{C}_x(s)d\mathbf{B}_s \quad (i = 1, \dots, n^*). \qquad (5.53)$$

Then we have

$$\mathbf{E}[\|\mathbf{r}_j^n\|^2] = \mathbf{E}\left[\left\|\int_{t_{i-1}^n}^{t_i^n} \boldsymbol{\mu}_x(s)ds\right\|^2\right] + 2\mathbf{E}\left[\left(\int_{t_{i-1}^n}^{t_i^n} \boldsymbol{\mu}_x(s)ds\right)' \mathbf{r}_i^*\right] + \mathbf{E}[\|\mathbf{r}_i^*\|^2],$$

and

$$\mathbf{E}\left[\left\|\int_{t_{i-1}^n}^{t_i^n} \mathbf{C}_x(s)d\mathbf{B}_s - \int_{t_{i-1}^n}^{t_i^n} \mathbf{C}_x(t_{i-1}^n)d\mathbf{B}_s\right\|^2\right] = O\left(\left(\frac{1}{n}\right)^2\right).$$

Then we can evaluate as

$$\mathbf{E}\left[\|\mathbf{r}_i^n\|^2 - \int_{t_{i-1}^n}^{t_i^n} \operatorname{tr}(\boldsymbol{\Sigma}_x(s))ds\right] = O\left(\left(\frac{1}{n}\right)^{3/2}\right). \qquad (5.54)$$

Hence we find that the effects of drift terms are negligible when estimating integrated volatility and the co-volatility function.

(On Stochastic Volatility) For each $s \in [t_j^n, t_{j+1}^n)$ $(j = 1, \dots, n-1)$, the differences of $\int_{t_{j-1}^n}^{t_j^n} \mathbf{C}_x(s)d\mathbf{B}(s) - \mathbf{C}_x(t_{j-1}^n)(B(t_j^n) - B(t_{j-1}^n))$ is stochastically negligible and $\mathbf{E}[\mathbf{C}_x(s)(B(t_j^n) - B(t_{j-1}^n))] = 0$ $(s > t_j)$. We use the fact that the main part of $\sqrt{m_n}(\hat{\sigma}_{gh}^{(x)} - \sigma_{gh}^{(x)})$ is the quadratic Brownian functional as (5.34).

Because we have the assumption of (3.37) and (3.38), the limits of $\sqrt{m_n}(\hat{\sigma}_{gh}^{(x)} - \sigma_{gh}^{(x)})$ are asymptotically and conditionally normal given the asymptotic covariances $\int_0^1 [\sigma_{gg}^{(x)}(s)\sigma_{hh}^{(x)}(s) + \sigma_{gh}^{(x)}(s)^2]ds$ for $g, h = 1, \ldots, p$. It is because we can apply the arguments for the stable convergence of discretized stochastic processes (Theorem 2.2.15 of Jacod and Protter (2012), for instance) to the present situation and hence we can extend the proof of Theorem 3.3 to the present case.   □

Some Remarks: As we have mentioned in Sect. 3.5, we need to use the stable convergence in Theorem 3.4 because the limiting integrated volatility and co-volatility functions can be random. The stable convergences in the general stochastic processes have been explained by Jacod (1997), Jacod and Shiryaev (2003) and Jacod and Protter (2012). As a more general reference on stable convergences, Hausler and Luschgy (2015) give the different approach to stable convergences including the discussion of the nested condition on filtrations and martingales.

References

Anderson, T.W. 1971. *The Statistical Analysis of Time Series*. New York: Wiley.

Anderson, T.W. 2003. *An Introduction to Statistical Multivariate Analysis*, 3rd ed. New York: Wiley.

Billingsley, P. 1995. *Probability and measure*, 3rd ed. New York: Wiley.

Hausler, E., and H. Luschgy. 2015. *Stable Convergence and Stable Limit Theorems*. Berlin: Springer.

Ikeda, N., and S. Watanabe. 1989. *Stochastic Differential Equations and Diffusion Processes*, 2nd ed. Amsterdam: North-Holland.

Jacod, J. 1997. On continuous conditional Gaussian martingales and stable convergence in law. *Séminaire de Probability*, vol. XXXVI, 383–401. Berlin: Springer.

Jacod, J., and P. Protter. 1998. Asymptotic error distributions for the euler method for stochastic differential equations. *The Annals of Probability* 26–1: 267–307.

Jacod, J., and P. Protter. 2012. *Discretization of Processes*. Berlin: Springer.

Jacod, J., and A.N. Shiryaev. 2003. *Limit Theorems for Stochastic Processes*. Berlin: Springer.

Chapter 6
Extensions and Robust Estimation (1)

Abstract We investigate the asymptotic properties of the SIML estimator and the micro-market price-adjustment mechanisms in the process of forming the observed transaction prices. We also investigate the problem of volatility estimation in the round-off error model, which is a nonlinear transformation model of hidden stochastic process.

6.1 Introduction

In this chapter and Chap. 7, we investigate the robustness of SIML estimation when we have the round-off error and the micro-market price adjustment mechanisms in the process of forming the observed transaction prices. We formulate the round-off error model as a nonlinear transformation of the underlying financial price process with micro-market noise. The statistical problem of the round-off error model of continuous stochastic processes has been investigated previously by Delattre and Jacod (1997), Rosenbaum (2009), and Li and Myckland (2012). However, our formulation has a new aspect and our motivation is the empirical observation that we have the tick-size effects (the minimum price change and the minimum order size) and we often observe bid-ask spreads on securities in actual financial markets.

Micro-market models including price adjustments have also been discussed in the economic micro-market literature (Engle and Sun 2007; Hansbrouck 2007 for instance). Among possible micro-market statistical models, we first take the (linear) price adjustment model proposed by Amihud and Mendelson (1987) as a benchmark case in our investigation. Then, we consider the linear and nonlinear price adjustment models in which a continuous martingale is the hidden intrinsic value on the underlying security. A new statistical feature of our approach is to utilize the nonlinear (discrete) transformations of the continuous-time diffusion process with discrete-time noise. We will utilize the formulation of relevant nonlinear time series models,

© The Author(s) 2018   59
N. Kunitomo et al., *Separating Information Maximum Likelihood Method
for High-Frequency Financial Data*, JSS Research Series in Statistics,
https://doi.org/10.1007/978-4-431-55930-6_6

namely the simultaneous switching autoregressive (SSAR) model due to Kunitomo and Sato (1999).

The main theme of this chapter is the fact that the observed price can differ from the underlying intrinsic value of the security. We can interpret this discrepancy by a nonlinear transformation from the intrinsic value to the observed price. We can represent the present situation as the nonlinear statistical model of an unobservable (continuous-time) state process and an observed (discrete time) stochastic process with measurement error. When the effects of measurement errors are present, the SIML estimator is robust; that is, it is consistent and asymptotically normal (or mixed-normal in the stable convergence sense) as the sample size increases under a set of assumptions. The required condition on the threshold parameter for the round-off errors is weak. The asymptotic robustness of the SIML method regarding integrated volatility and covariance has desirable properties over other estimation methods from large number of data for the underlying continuous stochastic process with micro-market noise. In Chap. 7, we investigate the multivariate problem when we need to estimate the covariance and the hedging coefficient.

6.2 A General Formulation

Let $y(t_i^n)$ be the ith observation of the (log-) price at t_i^n for $0 = t_0^n < t_1^n < \cdots < t_n^n = 1$. We consider the situation in which the underlying continuous-time stochastic process $X(t)$ $(0 \leq t \leq 1)$ is not necessarily the same as the observed (log-)price at t_i^n $(i = 1, \ldots, n)$ and is one-dimensional diffusion as

$$X(t) = X(0) + \int_0^t \mu_x(s)ds + \int_0^t \sigma_x(s)dB(s) \quad (0 \leq t \leq 1), \qquad (6.1)$$

where $\mu_x(s)$ is a predictable locally bounded drift term, $\sigma_x(s)$ $(= c_x(s))$ is an adapted continuous and bounded volatility process, and $B(s)$ is SBM. The statistical objective is to estimate the integrated volatility (or the quadratic variation)

$$\sigma_x^2 = \int_0^1 \sigma_x^2(s)ds \qquad (6.2)$$

of the underlying continuous process $X(t)$ $(0 \leq t \leq 1)$ from the set of observations on $y(t_i^n)$.

In this chapter, we consider the situation in which the observed (log-)price $y(t_i^n)$ is not necessarily a Brownian semi-martingale but is generated by

$$y(t_i^n) = h\left(X(t_i^n), y(t_{i-1}^n), u(t_i^n)\right), \qquad (6.3)$$

where $h(\cdot)$ is a measurable function. In (6.3), the (unobservable) continuous Brownian semi-martingale $X(t)$ $(0 \leq t \leq 1)$ is defined by (6.1), and $u(t_i^n)$ is the micro-market noise process. For simplicity, we assume that $u(t_i^n)$ are a sequence of independently and identically distributed random variables with $\mathbf{E}(u(t_i^n)) = 0$ and $\mathbf{E}(u(t_i^n)^2) = \sigma_u^2$ $(0 = t_0^n < t_1^n < \cdots < t_n^n = 1; t_i^n - t_{i-1}^n = 1/n, i = 1, \ldots, n)$.

There are special cases of (6.1) and (6.3) that can describe important aspects of modeling financial markets and high-frequency financial data for practical market applications. The simple (high-frequency) financial model with micro-market noise can be represented by

$$y(t_i^n) = X(t_i^n) + v(t_i^n), \tag{6.4}$$

where the underlying process $X(t)$ is given by (6.1) and $v(t_i^n)$ is a sequence of independently and identically distributed (i.i.d.) random variables.

The most important statistical aspect of (6.4) is the fact that it is an additive (signal-plus-noise) measurement-error model and the signal is a continuous process. However, there are economic reasons why the standard situation such as (6.4) is insufficient for applications. The round-off error models and the high-frequency financial models with micro-market price adjustment cannot be reduced to (6.4) but can be represented as special cases of (6.1) and (6.3).

6.3 The Basic Round-Off Error Model

First, we investigate the basic round-off error model with micro-market noise. One motivation for doing so is that in actual financial markets, transactions occur with the minimum tick size and the observed price data may not take a continuous path over time. For instance, the traded price and quantity usually have minimum size : The Nikkei-225 Futures, which have been the most important traded derivatives in Japan, have a minimum traded-size of 10 yen, while the Nikkei-225-stock index was around 9,000 yen in the year of 2011. (See Hansbrouck 2007 for details regarding the business practice of major stock markets in the USA) Hence, it is interesting and important to see the effects of round-off errors on the estimated integrated volatility.

Let $y_i = P(t_i^n)$ and

$$P(t_i^n) = g_\eta \left(X(t_i^n) + u(t_i^n) \right), \tag{6.5}$$

where the micro-market noise term $u(t_i^n)$ is a sequence of i.i.d. random variables with $\mathbf{E}[u(t_i^n)] = 0$ and $\mathbf{E}[u(t_i^n)^2] = \sigma_u^2$. The nonlinear function

$$g_\eta(x) = \eta \left[\frac{x}{\eta} \right] \tag{6.6}$$

is the round-off part of x, $[x]$ is the largest integer being equal or less than x, and the threshold parameter η is a (small) positive constant.

This model corresponds to the micro-market model with the restriction of a minimum price change with micro-market noise, and η is the parameter that sets minimum price change. In Fig. 6.1, we show typical sample paths of the standard round-off

Fig. 6.1 a:
Round-off-simulation-1 **b:**
Round-off-simulation-2

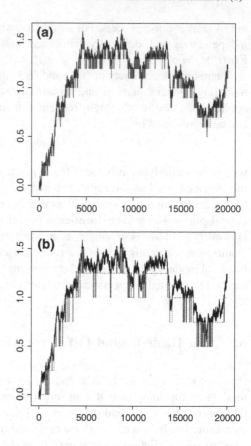

error model obtained from simulations. The true process is in black, while the observed process is the colored one with predetermined threshold level; e.g., $\eta = 0.1$, or 0.25.

6.4   A Micro-market Price Adjustment Model

Previously in financial economics, many micro-market models have been developed in attempts to explain the roles of noise traders, insiders, bid-ask spreads, transaction prices, the effects of taxes and fees, and the associated price adjustment processes in financial markets. As an illustration we give an underlying typical argument on the financial market mechanism by Fig. 6.2. In the underlying financial market, we let P be the price and Q be the quantity (in demand and supply) of a security. If the demand and supply curves of a security fail to meet, there is no transaction occurring at the same moment. The minimum (desired) supply price $\bar{P}$ is higher than the maximum (desired) demand price $\underline{P}$, whereupon there is a (bid-ask) spread. On the one hand,

Fig. 6.2 a: A micro-market model-1 **b**: A micro-market model-2

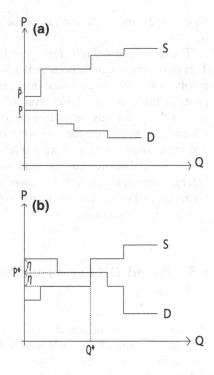

if there is some information on the supply side indicating that the intrinsic value of X_t of a security at time t is less than the latest observed price $P_{t-\Delta t}$ at time $t - \Delta t$ (i.e., $X_t - P_{t-\Delta t} < 0, \Delta t > 0$), the supply schedule is shifted downward. On the other hand, if there is some information on the demand side indicating that the intrinsic value of a security at time t is higher than the latest observed price (i.e., $X_t - P_{t-\Delta t} > 0$), the demand schedule is shifted upward. In these circumstances while the trade of a security occurs at price P^* and quantity Q^* as in Fig. 6.2, the financial market is under pressure for price changes. Figure 6.2a shows the case when there is no transaction, while Fig. 6.2b shows the case when there is a transaction and some excess demand remained.

Note: In Fig. 6.2, P and Q stand for the price and the quantity, respectively. D and S are the demand curve and supply curve, respectively. η in Fig. 6.2 denotes the minimum tick size, and Q^* is the quantity traded in Fig. 6.2.

Let $y(t_i^n) = P(t_i^n)$ $(i = 1, \ldots, n)$ and consider the (linear) micro-market price adjustment model given by

$$P(t_i^n) - P(t_{i-1}^n) = b\left(X(t_i^n) - P(t_{i-1}^n)\right) + u(t_i^n), \tag{6.7}$$

where $X(t)$ (the intrinsic log-value of a security at t) and $P(t_i^n)$ (the observed log-price at t_i^n) are measured logarithmically, the adjustment (constant) coefficient is b

$(0 < b < 2)$, and $u(t_i^n)$ is an i.i.d. noise sequence with $\mathbf{E}[u(t_i^n)] = 0$ and $\mathbf{E}[u(t_i^n)^2] = \sigma_u^2$.

The specific model is (6.7) proposed originally by Amihud and Mendelson (1987). This is a typical example because it has been one of well-known micro-market model involving transaction costs and interactions among different types of market participants in financial market. However, we depart from the Amihud–Mendelson model, and in that we focus on estimating integrated volatility whereas their main purpose was to investigate the micro-market mechanisms by using daily (open-to-open and close-to-close) data. Also, whereas Amihud and Mendelson (1987) had $X(t_i^n)$ following a (discrete) random walk in a discrete time series framework, we consider the case in which $X(t)$ is a general continuous-time Brownian semi-martingale given by (6.1), and the integrated volatility of the intrinsic value is defined by $0 < \int_0^t \sigma_s^2 ds < \infty$ (a.s.).

6.5   Round-Off Errors and Nonlinear Price Adjustment Models

We can generalize the round-off error model and the linear price-adjustment model we have introduced. We will discuss two variants for illustration, although there are others. As a nonlinear model, we represent the micro-market price-adjustment models with round-off error effects.

Let $y_i\ (= y(t_i^n)) = P(t_i^n)$ and

$$P(t_i^n) - P(t_{i-1}^n) = g_\eta \left(X(t_i^n) - P(t_{i-1}^n) + u(t_i^n) \right), \tag{6.8}$$

where $u(t_i^n)$ is a sequence of i.i.d. noises with $\mathbf{E}[u(t_i^n)] = 0$, $\mathbf{E}[u(t_i^n)^2] = \sigma_u^2$ and $g_\eta(x)$ is defined by (6.6). Then from (6.8), the difference between the observed price and the underlying intrinsic value can be represented as

$$\begin{aligned}
P(t_i^n) - X(t_i^n) &= g_\eta \left(-(P(t_{i-1}^n) - X(t_{i-1}^n)) + \Delta X(t_i^n) + u(t_i^n) \right) \\
&\quad + (P(t_{i-1}^n) - X(t_{i-1}^n) - \Delta X(t_i^n)) \\
&= g_\eta^* \left(P(t_{i-1}^n) - X(t_{i-1}^n), \Delta X(t_i^n), u(t_i^n) \right),
\end{aligned} \tag{6.9}$$

where

$$\Delta X(t_i^n) = \int_{t_{i-1}^n}^{t_i^n} \mu_x(s) ds + \int_{t_{i-1}^n}^{t_i^n} \sigma_x(s) d B_s$$

is a sequence of differences of $X(t_i^n)$ and $g_\eta^*(\cdot)$ is defined implicitly by (6.9). The round-off model of (6.8) and (6.9) can be represented as a nonlinear adjustment model with noise components.

Regarding the price-adjustment mechanism, there have been discussions on asymmetrical movements of financial price processes, which are related to problems in financial risk management. It may be natural to consider the situation in which there are different mechanisms in the upward phase of financial prices (i.e., $y(t_i^n) \geq y(t_{i-1}^n)$) and in the downward phase (i.e., $y(t_i^n) \leq y(t_{i-1}^n)$). In the context of micro-market models, some economists have tried to find appropriate econometric models, which involve transaction costs and micro-market structures with market practice and regulation on the limits of downward price movements within a day. As an example of discrete time series modeling of a nonlinear price of security price, we consider a nonlinear extension of (6.7) with

$$P(t_i^n) - P(t_{i-1}^n) = [b_1 I(X(t_i^n) \geq P(t_{i-1}^n)) + b_2 I(X(t_i^n) < P(t_{i-1}^n))][X(t_i^n) - P(t_{i-1}^n)] + u(t_i^n),$$
$$(6.10)$$

where b_i $(i = 1, 2)$ are some constants and $I(\cdot)$ is the indicator function.

This model is analogous to the class of SSAR models investigated by Kunitomo and Sato (1999), which is analogous to the TAR (threshold AR) models developed by Tong (1990) having delayed parameters. A set of sufficient conditions for the stability of the SSAR-type process is given by $b_1 > 0$, $b_2 > 0$ and $(1 - b_1)(1 - b_2) < 1$ with some moment condition on $u(t_i)$. If we set $b_1 = b_2 = b$, then we have the linear adjustment case as (6.7) and the stability condition is given by $0 < b < 2$. Nonlinear time series models such as the TAR model and the exponential AR model were developed in nonlinear time series analysis, and their statistical properties have been extensively discussed by Tong (1990).

To consider general nonlinear price adjustment models, that differ from the round-off-error models, we take $y_i = P(t_i^n)$ and

$$P(t_i^n) - P(t_{i-1}^n) = g\left(X(t_i^n) - P(t_{i-1}^n)\right) + u(t_i^n),$$
$$(6.11)$$

where $g(\cdot)$ is a nonlinear measurable function, and $u(t_i^n)$ is a sequence of i.i.d. micro-market noise with $\mathbf{E}[u(t_i^n)] = 0$ and $\mathbf{E}[u(t_i^n)^2] = \sigma_u^2$.

This representation includes the linear and nonlinear price adjustment modes as special cases.

6.6   Asymptotic Robustness of SIML for the Round-Off Error and Price Adjustment Models

We investigate the asymptotic properties of SIML estimation with the round-off error models and the micro-market adjustment models. First, we investigate the situation for the basic round-off model when we have a sequence of discrete observations $P(t_i^n)$ with $0 = t_0^n < t_1^n < \cdots < t_n^n = 1$ and we estimate the integrated volatility of the underlying security $\sigma_x^2 = \int_0^1 \sigma_x(s)^2 ds$.

Let $[x]$ and $\{x\}$ be the integer part and the fractional part of a real number x, respectively. If there is no micro-market noise term in (6.5), we can decompose

$$X(t_i^n) = \eta \left[\frac{X(t_i^n)}{\eta} \right] + \eta \left\{ \frac{X(t_i^n)}{\eta} \right\} \tag{6.12}$$

for $i = 1, \ldots, n$. Then, we have $y(t_i^n) = \eta[\frac{X(t_i^n)}{\eta}]$ and

$$\sum_{i=1}^{n} (\Delta y(t_i^n))^2 = \sum_{i=1}^{n} (\Delta X(t_i^n))^2 + \eta^2 \sum_{i=1}^{n} \left(\left\{ \frac{X(t_i^n)}{\eta} \right\} - \left\{ \frac{X(t_{i-1}^n)}{\eta} \right\} \right)^2$$

$$- 2\eta \sum_{i=1}^{n} (\Delta X(t_i^n)) \left(\left\{ \frac{X(t_i^n)}{\eta} \right\} - \left\{ \frac{X(t_{i-1}^n)}{\eta} \right\} \right),$$

where $\Delta y(t_i^n) = y(t_i^n) - y(t_{i-1}^n)$ and $\Delta X(t_i^n) = X(t_i^n) - X(t_{i-1}^n)$.

We set the threshold parameter $\eta = \eta_n$, which is dependent on n. If it satisfies the condition

$$\eta_n \sqrt{n} = o(1), \tag{6.13}$$

then the first term of the right-hand terms converges to the integrated volatility σ_x^2 as $n \to \infty$ because the realized volatility is a consistent estimator of the integrated volatility when we have no micro-market noise. However, if this condition is not satisfied and also there is a micro-market noise term at the same time, it is not obvious how to estimate the integrated volatility. Because n is finite in practice, we need to search for a weak condition on the threshold parameter η_n, and in this respect, we have the next result.

Theorem 6.1 *We assume that in (6.1) and (6.2), $\mu_x(s)$ and $\sigma_x(s)$ are bounded and deterministic functions of time, and further $\sigma_x(s)$ is bounded from zero. Also in (6.5) and (6.6), we assume that there exists γ ($\gamma > 0$) such that the threshold parameter $\eta = \eta_n$ satisfies*

$$\eta_n n^\gamma = O(1). \tag{6.14}$$

Define the SIML estimator of the integrated volatility of $X(t)$ with $m_n = [n^\alpha]$ ($0 < \alpha < 0.4$) by (3.18). Then, the limiting distribution of the normalized estimator $\sqrt{m_n} \left[\hat{\sigma}_x^2 - \sigma_x^2 \right]$ is asymptotically ($m_n, n \to \infty$) equivalent to the limiting distributions given by Theorem 3.1.

As second case, we consider the linear price adjustment model (6.7). The observed price at t_i^n ($i = 1, \ldots, n$) can be expressed as

$$P(t_i^n) = (1 - b)P(t_{i-1}^n) + bX(t_i^n) + u(t_i^n)$$

$$= b \sum_{j=0}^{i-1} (1 - b)^j X(t_{i-j}^n) + \sum_{j=0}^{i-1} (1 - b)^j u(t_{i-j}^n) + (1 - b)^i P(t_0^n),$$

which is a weighted linear combination of past intrinsic values and past noise terms. In this case, however, we notice from (6.7) that

$$P(t_i^n) - u(t_i^n) - X(t_i^n)$$
$$= (1 - b) \left[P(t_{i-1}^n) - X(t_i^n) \right]$$
$$= (1 - b) \left[P(t_{i-1}^n) - X(t_{i-1}^n) - u(t_{i-1}^n) \right]$$
$$+ (1 - b) \left[u(t_{i-1}^n) - \int_{t_{i-1}^n}^{t_i^n} \mu_x(s) ds - \int_{t_{i-1}^n}^{t_i^n} \sigma_x(s) d B_s \right]. \qquad (6.15)$$

We define a sequence of random variables $U_a(t_i^n) = P(t_i^n) - u(t_i^n) - X(t_i^n)$ and $W^*(t_i^n) = (1 - b)[u(t_{i-1}^n) - \int_{t_{i-1}^n}^{t_i^n} \mu_x(s) ds - \int_{t_{i-1}^n}^{t_i^n} \sigma_x(s) d B_s]$. Then, we can represent the present model as

$$U_a(t_i^n) = (1 - b) U_a(t_{i-1}^n) + W^*(t_i^n). \qquad (6.16)$$

Although $W^*(t_i^n)$ are correlated with $U_a(t_{i-1}^n)$, the linear price-adjustment model can be regarded as an extension of the basic model of (6.1) and (6.4). Then, we have the next result on the limiting distribution of the SIML estimator, the proof of which is given in the last subsection of this chapter.

Theorem 6.2 *We assume that $X(t)$ and u_i ($i = 1, \ldots, n$) in (6.1) and (6.7) are independent, and that b ($0 < b < 2$) in (6.7) is a constant. Define the SIML estimator of the integrated volatility of $X(t)$ with $m_n = [n^\alpha]$ ($0 < \alpha < 0.4$) by (3.18). Then, the asymptotic distribution of $\sqrt{m_n} \left[\hat{\sigma}_x^2 - \sigma_x^2 \right]$ is asymptotically ($m_n, n \to \infty$) equivalent to the limiting distributions given by Theorem 3.1.*

We notice that the present micro-market (linear) adjustment model is quite similar to the structure of the micro-market model with autocorrelated micro-market noises.

Third, we investigate the situation in which we have a sequence of price adjustments with the round-off error effect as (6.8) of Sect. 6.5.
Define

$$U_a(t_i^n) = P(t_i^n) - X(t_i^n) - u(t_i^n). \qquad (6.17)$$

Then, $U_a(t_i^n) = [P(t_i^n) - P(t_{i-1}^n)] + [P(t_{i-1}^n) - X(t_i^n) - u(t_i^n)]$ and the first term is $g_\eta(X(t_i^n) - P(t_{i-1}^n) + u(t_i^n))$. When $|P(t_{i-1}^n) - X(t_i^n) - u(t_i^n)| > \eta$, we use the fact that $x + \eta > x \geq g_\eta(x) > x - \eta$ and we have that $|U_a(t_i^n)| \leq \eta$. By contrast, when $|P(t_{i-1}^n) - X(t_i^n) - u(t_i^n)| \leq \eta$, then $P(t_i^n) = P(t_{i-1}^n)$ and $|U_a(t_i^n)| \leq \eta$.
Define $y_i = P(t_i^n)$ and $v_i = u(t_i^n) + U_a(t_i^n)$ ($i = 1, \ldots, n$), we have $y_i = x_i + v_i$ and

$$|U_a(t_i^n)| \leq \eta. \qquad (6.18)$$

By using similar arguments to the results reported as Theorem 3.1 on the limiting distribution of the integrated volatility estimator, we have the next result, the proof of which is given in the last subsection of this chapter.

Theorem 6.3 *We assume that $X(t)$ and u_i ($i = 1, \ldots, n$) in (6.1) and (6.8) are independent. The threshold parameter $\eta = \eta_n$ depends on n satisfying*

$$\eta_n \sqrt{n} = O(1) . \tag{6.19}$$

Define the SIML estimator of the integrated volatility of $X(t)$ with $m_n = [n^\alpha]$ ($0 < \alpha < 0.4$) by (3.18). Then, the limiting distribution of the normalized estimator $\sqrt{m_n}[\hat{\sigma}_x^2 - \sigma_x^2]$ is asymptotically ($m_n, n \to \infty$) equivalent to the limiting distributions given by Theorem 3.1.

In the above theorem, we have imposed the condition (6.19) on η, which is weaker than (6.13) but stronger than (6.14). This condition could be relaxed because our simulations suggest that the asymptotic result does not essentially depend upon these conditions.

Finally, we investigate the situation in which we have a sequence of discrete observations under the nonlinear adjustment model given by (6.11). We set $y(t_i^n) = P(t_i^n)$ and use a sequence of differences of $X(t_i^n)$ as $\Delta X(t_i^n) = X(t_i^n) - X(t_{i-1}^n) = \int_{t_{i-1}^n}^{t_i^n} \mu_x(s)ds + \int_{t_{i-1}^n}^{t_i^n} \sigma_x(s)dB_s$.

Also, we let

$$U_a(t_i^n) = P(t_i^n) - [X(t_i^n) + u(t_i^n)] \tag{6.20}$$

and

$$w(t_i^n) = -\Delta X(t_i^n) + u(t_{i-1}^n) . \tag{6.21}$$

The adjustment process $U_a(t_i^n)$, which is the difference between the observed price and the true process with noise, describes the additional price adjustment mechanism because $P(t_i^n) - X(t_i^n) = u(t_i^n) + U_a(t_i^n)$.

If there is no drift term (i.e., $\mu_x(s) = 0$), $w(t_i^n)$ are a sequence of uncorrelated random variables when $\Delta X(t_i^n)$ and $u(t_i^n)$ are independent. The difference between the observed price and the underlying intrinsic value plus noise can be represented as

$$\begin{aligned} U_a(t_i^n) &= U_a(t_{i-1}^n) + w(t_i^n) + g\left[-U_a(t_{i-1}^n) - w(t_i^n)\right] \\ &= g^*\left[U_a(t_{i-1}^n) + w(t_i^n)\right], \end{aligned} \tag{6.22}$$

where $g^*(z) = z + g(-z)$, $\mathbf{E}[w(t_i^n)] = 0$ and $\mathbf{E}[w(t_i^n)^2] < \infty$.

When we have an (ergodic) stationary solution for $U_a(t_i^n)$, we can reduce the model of (6.20)–(6.22) as a signal-plus-noise stationary process such that $y_i = x_i + v_i$ ($i = 1, \ldots, n$), where we set $y_i = P(t_i^n)$, $x_i = X(t_i^n)$ and $v_i = U_a(t_i^n) + u(t_i^n)$. However, by our construction we have the situation in which the signal term x_i and the noise term v_i are mutually correlated and v_i are autocorrelated over time. Because the discrete time series $U_a(t_i^n)$ satisfies a stochastic difference equation, it has a Markovian property.

To investigate the limiting behavior of the volatility estimation, we need a set of sufficient conditions, which are some type of ergodic condition. We summarize our results under some additional conditions with the nonlinear price adjustments, the proof of which is given in the last subsection of this chapter.

Theorem 6.4 *We assume that $X(t)$ does not have drift term and u_i $(i = 1, \ldots, n)$ in (6.1) and (6.11) are independent. For $U_a(t_i^n)$ satisfying (6.22) and $\mathbf{E}[U_a(t_i^n)^4] < \infty$, we further assume that there exist functions $\rho_1(\cdot)$ and $\rho_2(\cdot, \cdot)$ such that*

$$\text{Cov}[U_a(t_i^n), U_a(t_j^n)] = c_1\rho_1(|i - j|), \tag{6.23}$$

where c_1 is a (positive) constant, $\sum_{s=0}^{\infty} \rho_1(s) < \infty$ and

$$\text{Cov}\left[U_a(t_i^n)U_a(t_{i'}^n), U_a(t_j^n)U_a(t_{j'}^n)\right] = c_2\rho_2(i - i', j - j') \text{ for } j > j' > i > i',$$
$$\tag{6.24}$$

where c_2 is a (positive) constant and $\sum_{s,s'=0}^{\infty} \rho_2(s, s') < \infty$.

Define the SIML estimator of the realized volatility of $P(t_i^n)$ with $m_n = [n^\alpha]$ $(0 < \alpha < 0.4)$ by (3.18). Then, the asymptotic distribution of $\sqrt{m_n}\left[\hat{\sigma}_x^2 - \sigma_x^2\right]$ is asymptotically (as $m_n, n \to \infty$) equivalent to the limiting distributions given by Theorem 3.1.

The assumption of no drift in X is not essential, but a convenient one. See Theorem 3.4 of Chap. 3 on this problem.

In the above theorem, we impose a set of sufficient conditions for (6.24), which is a strong-mixing condition. If we can find positive constants ρ $(0 \le \rho < 1)$ and c_3 such that the conditional expectation

$$|\mathbf{E}[U_a(t_j^n)U_a(t_{j'}^n)|U_a(t_i^n), U_a(t_{i'}^n)]| < c_3\rho^{|j-j'|} \tag{6.25}$$

for any $j > j' > i > i'$, then we have (6.23) and (6.24).

There are many (discrete statistical) time series models satisfying the ergodicity conditions. A simple example is the linear case in which $g(z) = c\, z$ (c is a constant with $0 < c < 2$ and $U_a(t_i^n)$ represent a weakly dependent process. It is straightforward to have the above conditions in this case based on arguments that are similar to the derivations in Chap. 8 of Anderson (1971). The second example is a SSAR-type model by (6.11). If we impose a strong condition such as $0 < b_1, b_2 < 1$, it is also straightforward to use the arguments for linear processes. According to our simulations, however, this condition is often too strong to have the desired results and it may be an interesting future topic. There can be large number of nonlinear price-adjustment models and nonlinear discrete time series models for $X(t_i^n)$ and $P(t_i^n)$.

6.7 Simulations

We have investigated the robustness of the SIML estimator for integrated volatility based on a set of simulations, and the number of replications is 1,000. We took the sample size as $n = 20,000$, and chose $\alpha = 0.4$ (or 0.45) and $\beta = 0.8$ in all cases.

In our simulations, we consider several cases in which the observations are generated by (6.1) and (6.3). For simplicity, we set $\mu_x(s) = 0$ and the volatility function $(\sigma_x^2(s))$ is given by

$$\sigma_x^2(s) = \sigma(0)^2 \left[a_0 + a_1 s + a_2 s^2\right], \tag{6.26}$$

where a_i $(i = 0, 1, 2)$ are constants and we have some restrictions such that $\sigma_x(s)^2 > 0$ for $s \in [0, 1]$. This is a typical time-varying (but deterministic) case and the integrated volatility σ_x^2 is given by

$$\sigma_x^2 = \int_0^1 \sigma_x(s)^2 ds = \sigma_x(0)^2 \left[a_0 + \frac{a_1}{2} + \frac{a_2}{3}\right]. \tag{6.27}$$

In this example we take several intra-day volatility patterns including flat (or constant) volatility, monotone (decreasing or increasing) movements and U-shaped movements.

From many Monte-Carlo simulations, we summarize our main results in tabular form. We use several models in the form of (6.1) and (6.3), each of which corresponds to the typical cases when we take $h(\cdot, \cdot, \cdot)$ in (6.3) as

Model 1 $h_1(x, y, u) = \quad g_\eta(x + u)$ $(g_\eta(\cdot)$ is (6.6)) ,

Model 2 $h_2(x, y, u) = y + b(x - y) + u$ $(b :$ a constant) ,

Model 3 $h_3(x, y, u) = y + g_\eta(x - y + u)$ $(g_\eta(\cdot)$ is (6.6)) ,

Model 4 $h_4(x, y, u) = y + g_\eta(x - y) + u$ $(g_\eta(\cdot)$ is (6.6)) ,

Model 5 $h_5(x, y, u) = \quad y + u + \begin{cases} b_1(x - y) \text{ if } x \geq y \\ b_2(x - y) \text{ if } x < y \end{cases}$,

where b_i $(i = 1, 2)$ are constants.

Model 1 is the basic round-off error model in Sect. 6.2. Model 2 corresponds to the linear price adjustment model with the micro-market noise when the adjustment coefficient b is a constant. When $0 < b < 2$, Model 2 corresponds to the stationary linear price adjustment model with the micro-market noise. Models 1, 3, and 4 are micro-market models with round-off errors. Model 1 is the basic round-off error model, while Models 3 and 4 are the round-off error models with price adjustment mechanisms. Model 5 is the SSAR-type model, which is a typical nonlinear (discrete) time series model.

For comparison, we have calculated the historical integrated volatility (HI) estimate and the SIML estimate, which we compare in each table. Overall, the estimates by the SIML method are quite stable and robust against the possible values of the variance ratio even in the nonlinear transformations we have considered. For Model-1, the estimates obtained by historical-volatility (H-vol) are badly biased, which is a known situation in the analysis of high-frequency financial data. In fact, the values of H-vol are badly biased in all cases of our simulations, whereas the SIML method gives reasonable estimates in all cases of Models 1 to 5 (see Tables 6.1, 6.2, 6.3, 6.4, 6.5, 6.6, 6.7 and 6.8). We give some representative cases as Tables 6.3, 6.4, 6.5, 6.6,

Table 6.1 Comparison of alternative estimates (Model-1) ($\sigma_u^2 = 5.0\mathrm{E} - 05$, $\eta = 0.25$)

n = 20,000	SIML (σ_x^2)	H-vol	RK	PA
True-val	1.000	1.00	1.00	1.00
Mean	1.104	38.84	2.283	1.719
SD	0.258	9.354	0.514	0.387
MSE	0.092	1519.6	1.910	0.667

Table 6.2 Comparison of alternative estimates (Model-1) ($\sigma_u^2 = 1.0\mathrm{E} - 04$; $\eta = 0.1$)

n = 20,000	SIML (σ_x^2)	H-vol	RK	PA
True-val	1.000	1.00	1.00	1.00
Mean	0.969	12.14	0.948	0.934
SD	0.150	0.443	0.078	0.095
MSE	0.023	124.3	0.008	0.013

6.7 and 6.8, wherein in these tables the estimates by the SIML method are quite stable and robust against the possible values of the variance ratio even in the nonlinear transformations we considered. We have chosen several simulation results, which represent important aspects among many possibilities. Overall, the bias of the SIML estimator is often small and the variance of the SIML estimator is often stable.

We have compared the SIML estimates with alternative estimation methods in Tables 6.1 and 6.2. We have chosen two alternative well-known estimates, the realized kernel (RK) method and the pre-averaging (PA) method, which were developed by Barndorff-Nielsen et al. (2008) and Jacod et al. (2009), respectively. For a fair comparison, we tried to follow the recommendation by Barndorff-Nielsen et al. (2008) on the choice of kernel (Tukey–Hanning) and the bandwidth parameter H in the RK method and we took the triangular function $g(x)$, $\theta = 1$ and $K = \sqrt{n}$ in the PA method. An important issue in the RK method is the choice H, which depends on the noise variance and the instantaneous variance (which are in fact unknown); we choose $H = c\sqrt{\sigma_u^2/[\sigma_x^2/n]}$ although σ_u^2 and σ_x^2 are not known in advance. We have given two cases in the basic round-off model when the threshold parameter is relatively large ($\eta = 0.25$) and not so large ($\eta = 0.1$). It is interesting that the SIML method clearly dominates two methods in the first case, while three methods are comparable in the second case. It may be surprising to find that the SIML method gives reasonable estimates in all the cases. The biases of the RK method and the PA method can be relatively large, whereas the SIML method does not have much bias in some situations. Also, we find that the RK and PA estimation give reasonable estimates in some basic cases if we take reasonable value of the key parameters H, θ and K in many cases. The effects of not knowing the variance ratio may be more unfavorable to the RK method.

By examining these reported results and other simulations, we conclude that we can estimate the integrated volatility of the hidden martingale part reasonably by

Table 6.3 Estimation of integrated volatility (Model-2) ($a_0 = 1$, $a_1 = 0$, $a_2 = 0$; $\sigma_u^2 = 1.00\text{E} - 04$, $b = 0.2$)

n = 20,000	σ_x^2	H-vol
True-val	1.00E + 00	1.00E+00
Mean	1.01E+00	2.33E+00
SD	1.97E-01	2.32E-02
MSE	3.89E-02	1.78E+00

Table 6.4 Estimation of integrated volatility (Model-2) ($a_0 = 1$, $a_1 = 0$, $a_2 = 0$; $\sigma_u^2 = 1.00\text{E} - 05$, $b = 1.0$)

n = 20,000	σ_x^2	H-vol
True-val	1.00E+00	1.00E+00
Mean	9.88E-01	1.40E+00
SD	1.99E-01	1.40E-02
MSE	3.97E-02	1.60E-01

Table 6.5 Estimation of integrated volatility (Model-2) ($a_0 = 1$, $a_1 = 0$, $a_2 = 0$; $\sigma_u^2 = 1.00\text{E} - 06$, $b = 0.01$)

n = 20,000	σ_x^2	H-vol
True-val	1.00E+00	1.00E+00
Mean	8.40E-01	2.51E-02
SD	1.66E-01	5.41E-04
MSE	5.31E-02	9.50E-01

SIML estimation despite the possible nonlinear transformation such as the threshold models. When we have nonlinear transformations of the original unobservable security (intrinsic) values, however, the biases of the RK method and the PA method are not negligible in some cases.

Note: In the following tables, the estimates of the variances (σ_x^2) are calculated by the SIML method, whereas H-vol values are calculated by historical (or realized) volatility estimation. The term "true-val" means the true parameter value in simulations and mean, SD, and MSE correspond to the sample mean, the sample standard deviation, and the sample mean squared error of each estimator, respectively.

6.8 Derivation of Theorems

We give only the outlines of proofs because some parts are analogous to those in Sect. 6.5. We prove Theorems 6.3 and 6.1. This is because the first parts are similar.

Table 6.6 Estimation of integrated volatility (Model-3) ($a_0 = 7, a_1 = -12, a_2 = 6$; $\sigma_u^2 = 2.00\text{E} - 02, \eta = 0.5$)

n = 20,000	σ_x^2	H-vol
True-val	4.50E+01	4.50E+01
Mean	4.60E+01	1.37E+02
SD	1.05E+01	6.19E+00
MSE	1.11E+02	8.46E+03

Table 6.7 Estimation of integrated volatility (Model-4) ($a_0 = 1, a_1 = 0, a_2 = 0$; $\sigma_u^2 = 0.00\text{E} + 00$, $\eta = 0.005$)

n = 20,000	σ_x^2	H-vol
True-val	1.00E+00	1.00E+00
Mean	1.00E+00	6.85E-01
SD	1.94E-01	8.66E-03
MSE	3.77E-02	9.92E-02

Table 6.8 Estimation of integrated volatility (Model-5) ($a_0 = 1, a_1 = 0, a_2 = 0$; $\sigma_u^2 = 1.00\text{E} - 03$, $b_1 = 0.2, b_2 = 5$)

n = 20,000	σ_x^2	H-vol
True-val	1.00E+00	1.00E+00
Mean	1.02E+00	6.65E+01
SD	1.94E-01	1.66E+00
MSE	3.79E-02	4.30E+03

We give some additional arguments for Theorems 6.2 and 6.4. We use the notation K_i ($i \geq 1$) as positive constants.

Proof of Theorem 6.3 Most parts of this proof are very similar to the corresponding ones in the proof of Theorem 3.1 (or Theorem 3.3). We write $y_i = x_i + v_i$, $v_i = u_i + w_i$ ($i = 1, \ldots, n$), where $|w_i| \leq \eta_n$. Then, we need to check that the effects of a sequence of random variables w_i ($i = 1, \ldots, n$) are negligible under the additional assumption (6.19) with the threshold parameter η_n (> 0).

We illustrate the underlying arguments. As (5.21), from (6.4) and (6.5), we notice that

$$
\left[z_{kn}^{(2)} \right]^2 = n \left[\sum_{i=1}^{n} b_{ki} (u_i + w_i) \right]^2
$$

$$
= n \left[\sum_{i=1}^{n} b_{ki} u_i \right]^2 + 2n \left[\sum_{i=1}^{n} b_{ki} u_i \right] \left[\sum_{i=1}^{n} b_{ki} w_i \right] + n \left[\sum_{i=1}^{n} b_{ki} w_i \right]^2 . \quad (6.28)
$$

By using the Cauchy–Schwartz inequality under (6.19) and $n \sum_{j=1}^{n} b_{kj}^2 = a_{kn}$ ($k = 1, \ldots, n$), we have

$$\mathbf{E} \left[\sum_{i=1}^{n} b_{ki} w_i \right]^2 \le \eta_n^2 a_{kn} . \qquad (6.29)$$

Then, we can find a positive constant K_1 such that

$$\mathbf{E} \left[z_{kn}^{(2)} \right]^2 = n\mathbf{E} \left[\sum_{i=1}^{n} b_{ki}(u_i + w_i) \right]^2 \le K_1 a_{kn} \left[1 + \eta_n \sqrt{n} \right]^2 . \qquad (6.30)$$

Hence, under (6.19), the threshold effects in (6.29) and (6.30) are stochastically negligible. Then, we use similar arguments to other terms in the decomposition and we apply the same argument as the proof of Theorems 3.1 and 3.3.   □

Proof of Theorem 6.1 In the first part of the proof, we use the similar arguments to those for Theorem 6.3, but in the present situation it is possible to evaluate the expected value of (6.28) more precisely.

We use the fact that the Fourier series of x for any $0 < x < 1$ is given by

$$x = \frac{1}{2} - \sum_{s=1}^{\infty} \frac{\sin 2\pi s x}{\pi s} = \frac{1}{2} - \sum_{s=1}^{\infty} \left[\frac{e^{i2\pi s x} - e^{-i2\pi s x}}{2i\pi s} \right] . \qquad (6.31)$$

Then, except for the countable points of discontinuity, the fractional part $\{x\}$ of any real number x ($= [x] + \{x\}$) is given by (6.31) since $[x] = 0$. For a random variable X, let

$$\{X\}^* = \sum_{s=1}^{\infty} \frac{1}{2i\pi s} [(e^{i2\pi s X} - \mathbf{E}(e^{i2\pi s X})) - (e^{-i2\pi s X} - \mathbf{E}(e^{-i2\pi s X}))] . \qquad (6.32)$$

First, we assume that $u(t_i^n) = 0$ ($i = 1, \ldots, n$), $X(0) = 0$, $\mu_x(s) = 0$, and $\sigma_x(s)$ is a deterministic function such that $X_j = \int_0^{t_j^n} \sigma_x(s) dB_s$ is a Gaussian process. Then by using the Gaussianity, we find

$$\mathbf{E}(e^{i2\pi s X_j}) = \mathbf{E}(e^{-i2\pi s X_j}) = e^{-2\pi^2 s^2 \int_0^{t_j^n} \sigma_x(s)^2 ds} \qquad (6.33)$$

and for any j, k ($j > k$; $j, k = 1, \ldots, n$)

$$\mathbf{Cov}[\{x_j/\eta_n\}, \{x_k/\eta_n\}]$$

$$= \sum_{s,s'=1}^{\infty} \frac{-1}{4\pi^2 s s'} \mathbf{E}\{[(e^{i2\pi s x_j/\eta_n} - \mathbf{E}(e^{i2\pi s x_j/\eta_n})) - (e^{-i2\pi s x_j/\eta_n} - \mathbf{E}(e^{-i2\pi s x_j/\eta_n}))]$$

$$\times [(e^{i2\pi s' x_k/\eta_n} - \mathbf{E}(e^{i2\pi s' x_k/\eta_n})) - (e^{-i2\pi s' x_k/\eta_n} - \mathbf{E}(e^{-i2\pi s' x_k/\eta_n}))]\} .$$

Let $\mathscr{G}_{k,n}$ be the σ-field generated by the random variables X_l ($= X(t_l^n)$, $l = 1, \ldots, k$). Under the assumption that $0 < \sigma_* \leq \sigma_s \leq \sigma^*$, the conditional expected value for $j > k$ becomes

$$\mathbf{E}[(e^{i2\pi s X_j/\eta_n} - \mathbf{E}(e^{i2\pi s X_j/\eta_n})) - (e^{-i2\pi s X_j/\eta_n} - \mathbf{E}(e^{-i2\pi s X_j/\eta_n}))|\mathscr{G}_{k,n}]$$
$$\leq K_2 \, e^{-2\pi^2 s^2 \sigma_*^2 (j-k)/[n\eta_n^2]} \, ,$$

where K_2 is a positive constant. Because $e^{-x} < 1/x$ ($x > 0$), the sum converges to a finite value. In fact, under condition (6.14), for any $|j - k| > n^{\delta_1}$ ($0 < \delta_1 < 1$) we have $2\pi^2 s^2 \sigma_*^2 (j - k)/[n\eta_n^2] = O(n^{\delta_1 - 1 + 2\gamma})$, which goes to $+\infty$ as $n \to \infty$ if we can rake $2\gamma + \delta_1 > 1$. By setting $\delta_2 = \delta_1 - 1 + 2\gamma$, $-2\pi^2 s^2 \sigma_*^2 (j - k)/[n\eta_n^2] = O(n^{-\delta_2})$. Now we can evaluate

$$\mathbf{Var}\left[\eta_n \sum_{j=1}^n b_{kj} \left\{\frac{X_j}{\eta_n}\right\}^*\right] = \eta_n^2 \sum_{j,j'=1}^n b_{kj} b_{k,j'} \mathbf{Cov}\left[\left\{\frac{X_j}{\eta_n}\right\}^*, \left\{\frac{X_{j'}}{\eta_n}\right\}^*\right]. \quad (6.34)$$

We decompose summation (6.34) into two components as (a) $\sum_{|j-j'|<n^{\delta_3}}$, and (b) $\sum_{|j-j'|\geq n^{\delta_3}}$ for $0 < \delta_3 < 1$. Then, there are n^{δ_3} terms in (a) and there are $n - n^{\delta_3}$ terms, but which are of order $o(1/n)$. Thus, by using the relation (6.31) and Lemma 6.1 below, we can take constants K_3, K_4 and K_5 such that for $w_j = \eta_n\{X_j/\eta_n\}$ ($j = 1, \ldots, n$)

$$n\mathbf{E}\left[\sum_{j=1}^n b_{kj} w_i\right]^2 \leq K_9 \left[n \sum_{j=1}^n b_{kj}^2\right] \eta_n^2 \left[n^{\delta_3} + (n - n^{\delta_2})\frac{1}{n}\right]$$
$$= K_4 \, a_{kn}[n^\gamma \eta_n]^2 \leq K_5 \, a_{kn} \quad (6.35)$$

by taking $\gamma = \delta_3/2$ ($0 < \delta_3 < 1$) (we can take $\delta_1 = \delta_3$ for instance). Thus for the rest of evaluation, we can apply the arguments of the standard cases.

Next, we consider the case in which we have the micro-market noise term $u_i = u(t_j^n)$ ($j = 1, \ldots, n$). In the previous arguments, we replace $\eta_n\{(X_j + u_j)/\eta_n\}$ ($j = 1, \ldots, n$) instead of $\eta_n\{X_j/\eta_n\}$ and apply the same arguments. Because of the independence assumption on $X(t_j^n)$ and $u(t_j^n)$ ($j = 1, \ldots, n$), we can utilize the relation

$$\mathbf{E}[e^{i2\pi s(X_j+u_j)/\eta_n}] = \mathbf{E}[e^{i2\pi s X_j/\eta_n}]\mathbf{E}[e^{i2\pi s u_j/\eta_n}]. \quad (6.36)$$

When we have the drift term, the effects of $\mathbf{E}[|\int_{t_k^n}^{t_j^n} \mu_x(s)ds|^2] \leq K_6|t_j^n - t_k^n|$, which is asymptotically negligible. $\qquad\square$

By using straightforward calculations, we have the next relation.

Lemma 6.1 *For $j, k = 1, \ldots, n$, let $\theta_{jk} = [2\pi/(2n + 1)](j - 1/2)(k - 1/2)$ and $\theta_k = [2\pi/(2n + 1)](k - 1/2)$. Then,*

$$b_{kj} = \frac{2}{\sqrt{2n+1}}(\cos\theta_{kj} - \cos\theta_{k,j+1}) \qquad (6.37)$$

, and for any positive integer l $(l < n)$,

$$\sum_{j=l+1}^{n} b_{kj}b_{j,j-l} = 8\sin^2(\frac{\theta_k}{2})\cos(l\theta_k) + o(1) . \qquad (6.38)$$

Proof of Lemma 6.1 We use the representation

$$b_{kj} = \frac{1}{2n+1}\left[(1 - e^{i\theta_k})e^{i\theta_{kj}} + (1 - e^{-i\theta_k})e^{-i\theta_{kj}}\right] . \qquad (6.39)$$

Then, we have $\sum_{j=1}^{n} e^{i2\theta_{kj}} = 1/(1 - e^{i\theta_k})$ and $\sum_{j=1}^{n}(1 - e^{i\theta_k})^2 e^{i2\theta_{kj}} = 1 - e^{i\theta_k}$. Then, it is straightforward to show that for any integer l,

$$(2n+1)\sum_{j=1}^{n} b_{kj}b_{kj-l} = \sum_{j}\left\{e^{-il\theta_k}(1 - e^{i\theta_k})^2 e^{i2\theta_{kj}} + e^{il\theta_k}(1 - e^{-i\theta_k})^2 e^{-i2\theta_{kj}}\right.$$

$$\left. + ne^{il\theta_k}(1 - e^{i\theta_k})(1 - e^{-i\theta_k}) + ne^{-il\theta_k}(1 - e^{-i\theta_k})^2(1 - e^{i\theta_k})\right\}$$

$$\sim 4n\left[\sin^2\frac{\theta_k}{2}\right][2\cos(l\theta_k)]$$

when n is large.   □

Proof of Theorems 6.2 and 6.4 Most parts of the proof of Theorems 6.2 and 6.4 are essentially the same to the corresponding ones in the proof of Theorem 3.3. Hence, we give only the outline of our derivations. We define $v_i = V(t_i^n) + u(t_i^n)$ $(i = 1, \ldots, n)$ and write $y_i = x_i + v_i$, where $y_i = P(t_i^n)$ and $x_i = X(t_i^n)$ in (6.19). The essential difference is the presence of $U_a(t_i^n)$ terms, whereupon v_i $(i = 1, \ldots, n)$ are autocorrelated in the present situation.

By using conditions (6.23) and (6.24) and the Cauchy–Schwartz inequality, we find a positive constant K_7 such that

$$\mathbf{E}[z_{kn}^{(2)}]^2 = n\mathbf{E}\left[\sum_{i=1}^{n} b_{ki}v_i \sum_{j=1}^{n} b_{kj}v_j\right]$$

$$\leq n\sum_{s=0}^{n} c_1\rho_1(s)\left[\sum_{i=1}^{n} b_{ki}b_{k,i-s}\right]$$

$$\leq K_7 \times a_{kn} . \qquad (6.40)$$

We also evaluate the variance of

$$z_{kn}^{(2)2} - \mathbf{E}[z_{kn}^{(2)2}] = n \sum_{j,j'=1}^{n} b_{kj} b_{k,j'} \left[v_j v_{,j'} - \mathbf{E}(v_j v_{j'}) \right]$$

and the expectations of $\left[z_{kn}^{(2)2} - \mathbf{E}[z_{kn}^{(2)2}] \right] \left[z_{k',n}^{(2)2} - \mathbf{E}[z_{k',n}^{(2)2}] \right]$.

By using the condition imposed by (6.23) and (6.24), we can find a positive constant K_8 such that

$$n^2 \sum_{i,i'=1}^{n} \sum_{j,j'=1}^{n} b_{ki} b_{k,i'} b_{k',j} b_{k',j'} \rho_2(|i - i'|, |j - j'|) \sim K_8 \times a_{kn} a_{k',n} . \qquad (6.41)$$

Then by collecting each term, we can find a positive constant K_9 such that

$$\mathbf{E} \left[\frac{1}{\sqrt{m_n}} \sum_{j=1}^{m_n} (z_{kn}^{(2)2} - \mathbf{E}[z_{kn}^{(2)2}]) \right]^2 \leq \frac{K_9}{m_n} \sum_{k,k'=1}^{m_n} a_{kn} a_{k'n}$$

$$= O(\frac{1}{m_n} \times (\frac{m_n^3}{n})^2) , \qquad (6.42)$$

which is $O(m_n^5/n^2)$ because $\sum_{k=1}^{m} a_{kn} - O(m_n^3/n)$.

By taking care of these changes in our derivations, it is straightforward to prove Theorems 6.2 and 6.4 as for Theorem 3.3. $\qquad \square$

References

Amihud, Y., and H. Mendelson. 1987. Trading mechanisms and stock returns : An empirical investigation. *Journal of Finance* XLII (3): 533–553.

Anderson, T.W. 1971. *The Statistical Analysis of Time Series*. New York: Wiley.

Barndorff-Nielsen, O., P. Hansen, A. Lunde, and N. Shephard. 2008. Designing realized kernels to measure the ex-post variation of equity prices in the presence of noise. *Econometrica* 76 (6): 1481–1536.

Delattre, S., and J. Jacod. 1997. A central limit theorem for normalized functions of the increments of a diffusion process in the presense of round-off errors. *Bernoulli* 3–1: 1–28.

Engle, R., and Z. Sun. 2007. When is noise not noise: A micro-structure estimate of realized volatility, Working Paper.

Hansbrouck, J. 2007. *Empirical Market Microstructure*. Oxford: Oxford University Press.

Jacod, J., Yingying Li, Per A. Mykland, M. Podolskij and M. Vetter. 2009. Microstructure noise in the continuous case : The pre-averaging approach. *Stochastic processes and their applications* 119: 2249–2276.

Kunitomo, N., and S. Sato. 1999. Stationary and non-stationary simultaneous switching autoregressive models with an application to financial time series. *Japanese Economic Review* 50 (2): 161–190.

Li, and Myckland. 2012. Rounding Errors and Volatility Estimation. *Unpublished Manuscript.* Chicago University.

Rosenbaum, M. 2009. Integrated volatility and round-off error. *Bernoulli* 15–3: 687–720.

Tong, H. 1990. *Non-linear Time Series: A Dynamic System Approach.* Oxford: Oxford University Press.

Chapter 7
Extensions and Robust Estimation (2)

Abstract We further consider the asymptotic robustness of the SIML estimator under the micro-market price adjustment mechanisms in two-dimensional processes. In particular, we investigate the estimation problem of integrated volatility, covariance and the resulting hedging coefficient in the round-off error models, which is a nonlinear transformation of hidden process, and the price adjustment models. We also investigate the effects of random sampling observations.

7.1 Introduction

In this chapter, we investigate further the properties of the SIML estimation of integrated volatility, covariance, and the hedging coefficient when we have round-off error, micro-market noise, and randomly sampled data. Actual high-frequency financial data are recorded at random times, and the effects of this randomness could be significant when we have round-off error and micro-market noise. Empirically we have price adjustments mechanisms such as the minimum price change and the minimum order size and we observe bid-ask spreads in financial markets.

Also, non-synchronous sampling on the coviance estimation in high-frequency financial data is common. That is, actual transactions of two or more different financial commodities often occur at different high-frequency time periods. The problem of estimating covariance from non-synchronous data was first investigated by Hayashi and Yoshida (2005), who proposed the so-called Hayashi–Yoshida (H-Y) method for the situation in which there is no micro-market noise. There are several possible random sampling schemes for covariance estimation, and we shall adopt the refreshing scheme developed by Bandorff-Nielsen, et al. (2011). By synchronizing financial data, we consider the problem of estimating hedging coefficient and correlation coefficients when we have micro-market noise. These quantities have important roles in hedging risk and risk management. Because we must estimate the volatility as well as the covariances for this purpose, it may be difficult to use the H-Y method directly when we have micro-market noise.

© The Author(s) 2018   79
N. Kunitomo et al., *Separating Information Maximum Likelihood Method
for High-Frequency Financial Data*, JSS Research Series in Statistics,
https://doi.org/10.1007/978-4-431-55930-6_7

7.2   A Two-Dimensional Model

Let $y_s(t_i^s)$ be the ith observation of the (log-) price of the first asset at t_i^s for $0 = t_0^s < t_1^s < \cdots < t_{n_s^*}^s \leq 1$ and $y_f(t_j^f)$ be the j-th observation of the (log-) price of the second asset at t_j^f for $0 = t_0^f < t_1^f < \cdots < t_{n_f^*}^f \leq 1$, where $t_{n_s^*}^s = \max_{t_i^s \leq 1}\{t_i^s\}$, $t_{n_f^*}^f = \max_{t_i^f \leq 1}\{t_i^f\}$ and we denote n_a $(a = s, f)$ as constant indexes and n_a^* $(a = s, f)$ as (bounded) stochastic indexes.

We consider the situation in which the observed (log-)prices differ from the corresponding underlying continuous process $X_s(t)$ and $X_f(t)$ $(0 \leq t \leq 1)$, respectively. The two-dimensional continuous stochastic process $\mathbf{X}(t) = (X_s(t), X_f(t))'$ is a Brownian semi-martingale with

$$\mathbf{X}(t) = \mathbf{X}(0) + \int_0^t \boldsymbol{\mu}_x(s)ds + \int_0^t \mathbf{C}_x(s)d\mathbf{B}(s) \quad (0 \leq t \leq 1), \qquad (7.1)$$

where $\boldsymbol{\mu}_x(s)$ and $\mathbf{C}_x(s)$ are the 2×1 drift terms and the 2×2 volatility matrix, respectively, which are progressively measurable with respect to the σ-field $\mathscr{F}_t$ and $\mathbf{B}(s)$ is two-dimensional Brownian motion.

The first statistical objective is to estimate the quadratic variation or the integrated volatility matrix

$$\boldsymbol{\Sigma}_x = \int_0^1 \begin{pmatrix} \sigma_{ss}^{(x)}(r) & \sigma_{sf}^{(x)}(r) \\ \sigma_{sf}^{(x)}(r) & \sigma_{ff}^{(x)}(r) \end{pmatrix} dr = \begin{pmatrix} \sigma_{ss}^{(x)} & \sigma_{sf}^{(x)} \\ \sigma_{sf}^{(x)} & \sigma_{ff}^{(x)} \end{pmatrix} \qquad (7.2)$$

of the underlying continuous process $\mathbf{X}(t)$ $(0 \leq t \leq 1)$ from the set of discrete observations on $(y_s(t_i^s), y_f(t_j^f))$ with $i = 1, \ldots, n_s^*$ and $j = 1, \ldots, n_f^*$.

Assumption 7-I: The Brownian semi-martingale (7.1) satisfies the condition on drift and volatility terms such that $\boldsymbol{\mu}_x(s)$ and $\mathbf{C}_x(s)$ are continuous and bounded in $s \in [0, 1]$.

The basic high-frequency financial market model with micro-market noises can be represented by

$$y_s(t_i^s) = X_s(t_i^s) + v_s(t_i^s), \quad y_f(t_j^f) = X_f(t_j^f) + v_f(t_j^f), \qquad (7.3)$$

where the underlying process $X(t) = (X_s(t), X_f(t))'$ is the Brownian semi-martingale given by (7.1). In the basic model, we assume that $v_s(t_i^s)$ and $v_f(t_j^f)$ are a sequence of i.i.d. random variables with $\mathbf{E}(v_s(t_i^s)) = 0$, $\mathbf{E}(v_f(t_j^f)) = 0$, $\mathbf{E}(v_s(t_i^s)^2) = \sigma_{ss}^{(v)}$, $\mathbf{E}(v_f(t_j^f)^2) = \sigma_{ff}^{(v)}$, $\mathbf{E}(v_s(t_i^s)v_f(t_j^f)) = \delta(t_i^s, t_j^f)\sigma_{sf}^{(v)}$, where $\delta(\cdot, \cdot)$ is the indicator function.

With some notational complications, it is possible to extend the basic case with random sampling to several directions. First, we can consider the cases in which $v_s(t_i^s)$ and $v_f(t_j^f)$ are discrete stationary time series processes satisfying

$$v_s(t_i^s) = \sum_{j=0}^{\infty} \theta_j^s w_s(t_{i-j}^s) , \quad v_f(t_i^f) = \sum_{j=0}^{\infty} \theta_j^f w_f(t_{i-j}^f) , \tag{7.4}$$

where there exist ρ_a ($0 \le \rho_a < 1$; $a = s, f$) such that $\theta_j^a = O(\rho_a^j)$ and $w_s(t_i^a)$ ($a = s, f$) are a sequence of independent random variables with $\mathbf{E}(w_s(t_i^s)) = 0$, $\mathbf{E}(w_f(t_i^f)) = 0$, $\mathbf{E}(w_s(t_i^s)^2) < \infty$, and $\mathbf{E}(w_f(t_i^f)^2) < \infty$. We define the sequence of random variables $w_a(t_i^a) = 0$ for $t_i^a < 0$ ($a = s, f$), and we maintain the (weak) stationarity conditions of $v_a(t_i^a)$ in the MA representation (7.4) to simplify our arguments.

Second, an important aspect of (7.3) is the fact that it is an additive (signal-plus-noise) measurement error model. However, there are some reasons why the basic model (7.3) is insufficient for applications as discussed in Chap. 6. Then we shall further consider several examples of the more general situation when the observed (log-)prices $y_a(t_i^a)$ are the sequence of discrete stochastic processes generated by

$$y_a(t_i^a) = h_a\left(\mathbf{X}(t_i^a), y_a(t_{i-1}^a), u_a(t_i^a)\right) \quad (a = s, f), \tag{7.5}$$

where $h_a(\,\cdot\,)$ are measurable functions, the (unobservable) continuous martingale process $\mathbf{X}(t)$ ($0 \le t \le 1$) is defined by (7.3), and the micro-market noise $u_a(t_i^a)$ and $u_f(t_i^f)$ are discrete stochastic processes.

We are interested in special cases in the form of (7.1) and (7.5), which reflect the important aspects regarding modeling financial markets and high-frequency financial data. As a nonlinear transformation, we consider the case in which $y_a(t_i^n) = P_a(t_i^n)$ and

$$P_a(t_i^a) = g_{a,\eta}\left[X_a(t_i^a) + u_a(t_i^a)\right] , \tag{7.6}$$

where the micro-market noise term $u_a(t_i^a)$ are sequences of i.i.d. random variables with $\mathbf{E}[u_a(t_i^a)] = 0$, $\mathbf{E}[u_a(t_i^a)^2] = \sigma_{a,u}^2$, the nonlinear function

$$g_{a,\eta}(x) = \eta_a \left[\frac{x}{\eta_a}\right] \tag{7.7}$$

is the round-off part of x, $[x]$ is the largest integer equal to or less than x, and the η_a are (small) positive constants.

Another direction may be the (linear) micro-market price adjustment model, which is given by $y_a(t_i^a) = P_a(t_i^a)$ ($i = 1, \ldots, n_a^*$; $a = s, f$) and

$$P_a(t_i^a) - P_a(t_{i-1}^a) = b_a\left[X_a(t_i^a) - P_a(t_{i-1}^a)\right] + u_a(t_a^a) , \tag{7.8}$$

where $X_a(t)$ (the intrinsic value of a security at time t) and $P_a(t_i^a)$ are the observed log-price at t_i^a, the adjustment (constant) coefficients are b_a ($0 < b_a < 2$), and $u_a(t_i^a)$ is an i.i.d. noise sequence with $\mathbf{E}[u(t_i^a)] = 0$ and $\mathbf{E}[u(t_i^a)^2] = \sigma_{aa}^2$.

Also, if we set $y_a(t_i^a) = P_a(t_i^a)$ and

$$P_a(t_i^a) - P_a(t_{i-1}^a) = g_{\eta,a}\left[X_a(t_i^a) - P_a(t_{i-1}^a) + u_a(t_i^a)\right], \qquad (7.9)$$

where the round-off error function $g_{a,\eta}(x)$ is defined by (7.7), and $u_a(t_i^a)$ is a sequence of i.i.d. noise with $\mathbf{E}[u_a(t_i^a)] = 0$ and $\mathbf{E}[u_a(t_i^a)^2] = \sigma_{aa}^2$.

The basic (high-frequency) financial model with micro-market noise is a special case when the underlying process $X(t) = (X_s(t), X_f(t))'$ is given by (7.1). The synchronous sampling means $t_i^s = t_i^f$, and the fixed grid observation means $t_i^a - t_{i-1}^a = n^{-1}$. There can be special cases of (7.5) such as (7.6), (7.8), and (7.9) when we have micro-market adjustment models and non-synchronous observations as well as random sampling. In this chapter, we consider the situation in which the high-frequency data are observed at random time t_i^a $(a = s$ or $f)$ under some conditions on random sampling. There are several ways to handle the problem of non-synchronously observed data for covariance estimation. Hayashi and Yoshida (2005), for instance, introduced the Hayashi–Yoshida method of covariance estimation, which is given by

$$\mathrm{HY}(s, f) = \sum_{i,j} [y_s(t_i^s) - y_s(t_{i-1}^s)][y_f(t_j^f) - y_f(t_{j-1}^f)]\mathbf{I}[(t_i^s - t_{i-1}^s) \cap (t_j^f - t_{j-1}^f) \neq \phi].$$

$$(7.10)$$

In this chapter, however, we mainly adopt the refreshing time method developed by Harris et al. (1995) and used by Bandorff-Nielsen, et al. (2011). Define $0 = t_0, t_1 = \max\{t_1^s, t_1^f\}$ and $t_{j+1}^n = \max\{t_{N_j^s+1}^s, t_{N_j^f+1}\}$, where $t_{N_j^a+1}^a$ are random times for $y_a(t_j^a)$ $(a = s, f)$, and N_j^a is the corresponding counting process. We denote the resulting random times as $0 = t_0 < t_1 < \cdots < t_{n^*}$ and the random number of observations as n^*. Then the resulting counting process N_j^n and n^* are finite-valued random variables in $[0, 1]$ for any finite n, and we denote $t_{n^*}^n$ (≤ 1) is the last transaction time before the market closing time in a day. We sometimes use the notations n and n^* for n_a and n_a^* $(a = s, f)$, respectively, for the sake of notational simplicity in the following analysis and statements whenever there is no cause of confusion.

Assumption 7-II: The sampling process $\{t_j^n\}$ is independent of the underlying process $\{X(t)\}$, and n^* is a finite-valued random variable in $[0, 1]$ for any n. There exist positive constants c_s, c_f, c and an increasing sequence of fixed n such that as $n \to \infty$

$$t_{n^*} \xrightarrow{p} 1 \,, \quad \frac{n_a^*}{n} \xrightarrow{p} c_a \ (a = s \text{ or } f) \,, \quad \frac{n^*}{n} \xrightarrow{p} c \,. \qquad (7.11)$$

Let $\Delta_j^n t_j^n = t_j^n - t_{j-1}^n = (1/n)D_j^n$ be a sequence of $\mathscr{F}_{n,j-1}$–adapted random variables. The bounded increasing continuous process $\tau(t)$ with $\tau(0) = 0, \tau(1) = 1$ and the continuous process $d(t)$ $(0 \leq t \leq 1)$ are well-defined such that $t_j^n = \tau(t) + O_p(n^{-\gamma_1})$ $(\gamma_1 > 0)$ and $D_j^n = d(t) + O_p(n^{-\gamma_2})$ $(\gamma_2 > 0)$ uniformly in t $(t \in (0, 1])$ as $j(n)/n \to t$ and $n \to \infty$.

These conditions imply that $t_j^n \xrightarrow{p} \tau(t)$, $D_j^n \xrightarrow{p} d(t)$ and $[D_j^n(t)]^2 \xrightarrow{p} d(t)^2$ uniformly (as $j(n)/n \to t$ and $n \to \infty$), where $\mathbf{E}\left[|t_i^n - t_{i-1}^n|\right] = O(n^{-1})$ is proportional to the average duration on intervals in $[0, 1]$ for any fixed n. For the standard normalization, we often take $c_s = c_f = 1$ or $c = 1$ in the following analysis. A typical example of random sampling is Poisson sampling on t_i^a with the intensity functions $\lambda_n^{(a)} = nc_a$ $(a = s, f)$. In this case, the (mutually) independent random durations D_i^a $(a = s, f)$ are exponentially distributed with $\mathbf{E}(D_i^a) = 1$, $nD_i^a = t_i^a - t_{i-1}^a$ and $\tau(t) = t$ $(0 < t \leq 1)$ if we take the normalization $c = 1$.

Integrated Hedging Ratio and Correlation

We illustrated the use of the hedging coefficient in Chap. 4. For financial risk-managements, the role of hedging coefficient and correlation coefficient has often been discussed in the literature on financial futures (see (Duffie 1989) for instance). Then it is important to estimate the hedging ratio and the correlation coefficient from a set of discrete sampled price or log-price data. The (integrated) hedging ratio based on high-frequency financial data can be defined by

$$H = \frac{\sigma_{sf}^{(x)}}{\sigma_{ff}^{(x)}} . \tag{7.12}$$

The (integrated) correlation coefficient between two prices can be defined by

$$\rho_{sf} = \frac{\sigma_{sf}^{(x)}}{\sqrt{\sigma_{ss}^{(x)} \sigma_{ff}^{(x)}}} . \tag{7.13}$$

7.3 Asymptotic Properties of SIML Estimation

It is important to investigate the asymptotic properties of the SIML estimator when the volatility function $\Sigma_x(s)$ is not constant over time. The asymptotic properties of the SIML estimator were given in Chap. 3. We give a slightly general result as the next proposition when the observations are randomly sampled in this section. For the case of stochastic volatility and covariance in continuous-time, we assume that $\mathbf{C}_x(t) = (c_{ij}^{(x)})$ (2×2) follows

$$c_{ij}^{(x)}(t) = c_{ij}^{(x)}(0) + \int_0^t \mu_{ij}^\sigma(s)ds + \int_0^t \omega_{ij}^\sigma(s)d\mathbf{B}^\sigma(s) , \tag{7.14}$$

where the drift coefficients $\mu_{ij}^\sigma(s)$ and diffusion vectors (1×2) $\omega_{ij}^\sigma(s)$ of the volatilities are extensively measurable, continuous, and bounded, and $\mathbf{B}^\sigma(s)$ is a 2×1 Brownian motion vector which can be correlated with $\mathbf{B}(s)$.

The asymptotic properties of the SIML estimator when data are randomly sampled can be summarized as Theorem 7.1, the proof of which is given in the last section of this chapter. We often use the notations n and n^* for n_a and n_a^* $(a = s, f)$, respectively, whenever no confusion arising from doing so.

Theorem 7.1 *We assume that $X(t)$ and $v_a(t_i^a)$ ($a = s, f, i, j \geq 1$) in (7.1) and (7.3) are independent and $v_a(t_i^a)$ are sequences of independent ranfom variables with $\mathbf{E}[v_s(t_i^s)] = 0$ and $\mathbf{E}[v_f(t_j^f)^2] < \infty$. Also, we assume the conditions on $\mathbf{C}_x(s)$ in (7.14) and $\mathbf{\Sigma}_x$ in (7.2) is nonnegative definite, and we involke Assumptions 7-I and 7-II with $c = 1$ as the normalization. We further assume that $\mathbf{E}[v_s(t_i^s)^4] < \infty$, $\mathbf{E}[v_f(t_j^f)^4] < \infty$.*

(i) For $m_n = [n^\alpha]$ and $0 < \alpha < 0.5$, as $n \longrightarrow \infty$, $m_{n^*}/m_n \overset{P}{\to} 1$,

$$\hat{\sigma}_{ss}^{(x)} - \sigma_{ss}^{(x)} \overset{P}{\longrightarrow} 0 \, , \; \hat{\sigma}_{ff}^{(x)} - \sigma_{ff}^{(x)} \overset{P}{\longrightarrow} 0 \tag{7.15}$$

and

$$\hat{\sigma}_{sf}^{(x)} - \sigma_{sf}^{(x)} \overset{P}{\longrightarrow} 0 \, . \tag{7.16}$$

(ii) For m_{n^*} and $0 < \alpha < 0.5$, as $n \longrightarrow \infty$, $m_{n^*}/m_n \overset{P}{\to} 1$,

$$\sqrt{m_n} \left[\hat{\sigma}_{ss}^{(x)} - \sigma_{ss}^{(x)} \right] \overset{\mathscr{L}-s}{\longrightarrow} N\left[0, V_{ss}\right] \, , \; \sqrt{m_n} \left[\hat{\sigma}_{ff}^{(x)} - \sigma_{ff}^{(x)} \right] \overset{\mathscr{L}-s}{\longrightarrow} N\left[0, V_{ff}\right] \, , \tag{7.17}$$

and

$$\sqrt{m_n} \left[\hat{\sigma}_{sf}^{(x)} - \sigma_{sf}^{(x)} \right] \overset{\mathscr{L}-s}{\longrightarrow} N\left[0, V_{sf}\right] \, , \tag{7.18}$$

where $Z_n \overset{\mathscr{L}-s}{\longrightarrow} Z$ is the stable convergence in law with

$$V_{ss} = 2 \int_0^1 \left[\sigma_{ss}^{(x)}(\tau(s)) \right]^2 d(s)ds \, , \; V_{ff} = 2 \int_0^1 \left[\sigma_{ff}^{(x)}(\tau(s)) \right]^2 d(t)ds \tag{7.19}$$

and

$$V_{sf} = \int_0^1 \left[\sigma_{ss}^{(x)}(\tau(s)) \sigma_{ff}^{(x)}(\tau(s)) + (\sigma_{sf}^{(x)}(\tau(s)))^2 \right] d(s)ds \, . \tag{7.20}$$

We make some remarks on the conditions we have used in Theorem 7.1 and some possibilities of extending them.

First, we have given Theorems 3.1, 3.3, and 3.4 for the basic case, the time-changing deterministic case, and the time-changing stochastic case of volatility in Chap. 3, respectively. We have presented the basic result for these cases with stochastic sampling. It may be possible to extend the results to more general cases such as the round-off error and micro-market price adjustment models we mentioned in Chap. 6. Nonetheless, we only show some simulation results in Sect. 7.4 for these models.

Second, when $v_s(t_i^s)$, $v_f(t_j^f)$ ($i, j \geq 1$) are autocorrelated and have the representation (7.4) with the fourth-order moment conditions that $\mathbf{E}[w_s(t_i^s)^4] < \infty$ and $\mathbf{E}[w_f(t_j^f)^4] < \infty$, we can strengthen the same result in Theorem 7.1 with $0 < \alpha < 0.4$.

Third, when the variance–covariance matrix is constant, the number of observations is constant as $n^* = n$, whereupon $t_i^s - t_{i-1}^s = 1/n, t_j^f - t_{j-1}^f = 1/n$ ($i, j =$

$1, \ldots, n)$ and $\tau(t) = t, d(t) = 1, \tau(t) - \tau(s) = t - s$ for $0 < s < t < 1, \sigma_{ss}^{(x)}(s) = \sigma_{ss}^{(x)}, \sigma_{ff}^{(x)}(s) = \sigma_{ff}^{(x)}, \sigma_{sf}^{(x)}(s) = \sigma_{sf}^{(x)}$, $\tau(t) = t, d(t) = 1, \tau(t) - \tau(s) = t - s$ for $0 \leq s < t \leq 1$. The asymptotic variance and covariance are given by

$$V_{ss} = 2[\sigma_{ss}^{(x)}]^2 , \quad V_{sf} = \sigma_{ss}^{(x)}\sigma_{ff}^{(x)} + [\sigma_{sf}^{(x)}]^2 . \tag{7.21}$$

There can be several cases in which we have the representation of (7.5) and (7.6) instead of (7.3) we have mentioned. We can extend the asymptotic results further, but we omit the details in this chapter.

By using the estimators of integrated volatility and integrated covariance, the SIML estimator of the hedging ratio $H = \sigma_{sf}^{(x)}/\sigma_{ff}^{(x)}$ is defined by (7.12). From Theorem 7.1, we use the standard delta method to approximate the normalized hedging estimator around the true coefficient as

$$\sqrt{m_n}(\hat{H} - H) \sim \frac{1}{\sigma_{ff}^{(x)}} \sqrt{m_n}(\hat{\sigma}_{sf}^{(x)} - \sigma_{sf}^{(x)}) - \frac{\sigma_{fs}^{(x)}}{(\sigma_{ff}^{(x)})^2} \sqrt{m_n}(\hat{\sigma}_{ff}^{(x)} - \sigma_{ff}^{(x)}) .$$

Then by evaluating the asymptotic variances and covariance $\mathbf{V}(\hat{\sigma}_{fs}^{(x)})$, $\mathbf{V}(\hat{\sigma}_{ff}^{(x)})$ and $\mathbf{Cov}(\hat{\sigma}_{ff}^{(x)}, \hat{\sigma}_{fs}^{(x)})$ (see Theorems 3.3 and 7.1), the resulting asymptotic variance as

$$V[\hat{H}] = \left[\frac{1}{\sigma_{ff}^{(x)}}\right]^2 V_{sf} + \left[\frac{\sigma_{sf}^{(x)}}{\sigma_{ff}^{(x)2}}\right]^2 V_{ff} - 4\frac{\sigma_{sf}^{(x)}}{\sigma_{ff}^{(x)3}} \int_0^1 \sigma_{sf}^{(x)}(\tau(s))\sigma_{ff}^{(x)}(\tau(s))d(s)ds .$$

$$\tag{7.22}$$

Although this formula looks complicated, when the volatility and covariance functions are constant in the basic case with the fixed (non-random) intervals, we have the next result.

Corollary 7.1 *Assume that the instantaneous volatility matrix* $\mathbf{C}_x(s)$ *and* $\mathbf{\Sigma}_x$ *are constant in the basic case with the fixed (non-random) intervals. Then the asymptotic variance of the limiting distribution of* $\sqrt{m_n}\left[\hat{H} - H\right]$ *is given by*

$$\omega_H = \frac{\sigma_{ss}^{(x)}}{\sigma_{ff}^{(x)}} \left[1 - \frac{\sigma_{sf}^{(x)2}}{\sigma_{ss}^{(x)}\sigma_{ff}^{(x)}}\right] . \tag{7.23}$$

7.4 Further Simulations

We have conducted large number of simulations. For each table, we have calculated the historical volatility (the realized volatility, RV), the realized co-volatiltity (RCV), the Hayashi–Yoshida (HY) estimator, and the SIML estimator. In tables, Raw means the estimates based on all simulated data and 10s means the estimates based on the simulated data at each grid of 10 s. We used the round-off error and micro-market price adjustment models which correspond to those in Sect. 7.2. From many Monte Carlo simulations, we summarize our main results in Tables 7.1, 7.2, 7.3, 7.4, and 7.5.

In Table 7.2, we use the EACD(1,1) model, as proposed originally by Engle and Russell (1998). Financial econometricians are interested in the exponential autoregressive conditional duration (EACD) models because the assumption of Poisson random sampling leads to a sequence of i.i.d. random variables for durations whereas we may have the dependent structure on the observed durations, which are length of execution times of traded prices. Although there are many types of duration dependent models, we use the EACD(1,1) model as a representative one. For $a = s$ or f, let $\tau_i^a = t_i^a - t_{i-1}^a$ and $\tau_i^a = \psi_i^a \varepsilon_i^a$, where

$$\psi_i^a = \omega + \delta \varepsilon_i^a + \gamma \psi_{i-1}^a \tag{7.24}$$

and ε_i^a are a sequence of i.i.d. exponential random variables with $\delta > 0, \gamma > 0, \omega > 0$ $E[\varepsilon_i^a]=0$ and $V[\varepsilon_i^a]=1$. We set $\delta=0.06$, $\gamma=0.9$ and $\omega = 0.04$ because $\omega = $ (average duration)$\times[1 - (\delta + \gamma)]$ when $n \sim 1800$ and $n \sim 18,000$ with the standardization of 1s. The number of replications is 1000 in all cases.

In our simulations, we use several nonlinear transformation models in the form of (7.5). We take $X_s(t_i^s)$ and $X_f(t_j^f)$ individually and each model in our simulation corresponds to

Model 1 $h_{a,1}(x, y, u) = x + u$,

Model 2 $h_{a,2}(x, y, u) = g_{a,\eta}(x + u)$ (where $g_{a,\eta}(z) = \eta_a[z/\eta_a]$ (7.7)) ,

Model 3 $h_{a,3}(x, y, u) = y + b_a(x - y) + u$ (b_a : constants) ,

Model 4 $h_{a,4}(x, y, u) = y + g_{\eta_a,a}(x - y + u)$
                    (where $g_{a,\eta}(z) = \eta_a[z/\eta_a]$ (7.7)) ,

Model 5 $h_{a,5}(x, y, u) = y + u + \begin{cases} b_{1,a}(x - y) \text{ if } x \geq y \\ b_{2,a}(x - y) \text{ if } x < y \end{cases}$,

where $b_{i,a}$ ($i = 1, 2; a = s, f$) are constants.

For the sake of simplicity, in our simulations for $a = s, f$, we set $\lambda_a = \lambda$ (Poisson intensity), $b_a = b$, $b_{i,a} = b_i$ ($i = 1, 2$) and the same values for parameters in two-dimensional process. We also standardized the value of λ and other parameters such that we have approximately $n \sim 1,800$ and $n \sim 18,000$ (i.e., $c_a \sim 1$ and $c \sim 1$ in Assumption 7-II), we only give the case of $n \sim 18,000$ in this chapter. In Table 7.1, the average and standard deviations are 17,995 and 139 in actual 1000 replications, respectively, for instance.

Model 1 is the basic (signal-plus-noise) additive model and Model 2 is the basic round-off error model. Model 3 corresponds to the linear price adjustment with micro-market noise and when the adjustment coefficients $0 < b_a < 2$, which is a sufficient condition for the corresponding adjustment mechanism to be ergodic. Model 4 is the micro-market model with round-off error and price adjustments, and Model 5 is the micro-market noise model with asymmetric nonlinear price adjustment.

When there are micro-market noise components with the martingale signal part, the values of realized variance (RV) and realized covariance (RCV) often differ substantially from the true integrated variance and covariance of the signal part. However, we have found that it is possible to use SIML estimation for the integrated volatilities, integrated covariance, and noise variances when we have the signal-to-noise ratio as $10^{-2} \sim 10^{-6}$. Although we have omitted the details, the estimation results are similar in the stochastic volatility model.

Table 7.1 gives the simulation results for the basic round-off error models with Poisson sampling and constant intensity λ. Since there are two variables, we say σ_{xi}^2 ($i = 1, 2$) and σ_{vi}^2 ($i = 1, 2$) are the variances of each intrinsic variables and noises. RV_i ($i = 1, 2$) are the realized volatilities, and RCV corresponds to the realized covariance. To construct the estimator of hedging coefficient, there can be several combinations of the covariance and variance estimates, and we have denoted RCV-RC, HY-RV, and SIML-SIML. As we showed in Chap. 6, the SIML estimator may dominate the existing methods including the realized kernel method and the pre-average method in some round-off error cases. Selected simulation results for other linear and nonlinear price adjustment models are given in Tables 7.3, 7.4, and 7.5.

For the covariance estimation, we have confirmed that using the H-Y method improves the historical covariance method as was pointed out by Hayashi and Yoshida (2005) in the absence of micro-market noise. However, in the presence of micro-market noise, the volatility estimation bias of the H-Y method can be large, whereas SIML estimation gives reasonable estimates of the hedging coefficient. There are alternative ways to use the integrated volatility and integrated covariance for estimating the hedging coefficient, but the SIML-SIML combination (i.e., SIML volatility estimate and SIML covariance estimate) gives reasonable estimates in all cases that we investigated. This point is vivid and important for practical risk management purposes, from Tables 7.1, 7.2, 7.3, 7.4, and 7.5.

By examining these results of our simulations in addition to the basic cases in Chap. 4, we conclude that we can estimate the integrated volatility and integrated covariance of the hidden continuous part reasonably well by SIML estimation. It may be surprising to find that the SIML method gives reasonable estimates even when we have nonlinear transformations of the original unobservable security (intrinsic) values. We conducted large number of further simulations, but the results are essentially the same to those reported in this section.

7.5 On Mathematical Derivations

In this subsection, we give some details of the proof of Theorem 7.1. The method of our derivations is essentially the same to the derivations reported in Chaps. 5 and 6, but we need some extra arguments. We denote K_i ($i \geq 1$) as positive constants.

Table 7.1 Estimation of covariance and hedging coefficient : Model 2 ($n \sim 18,000$; $\eta = 0.001$)

18000	True	Raw	1 s	10 s	300 s
σ_{x1}^2	2.00E-04	2.04E-04	2.05E-04	2.06E-04	2.10E-04
		4.20E-05	4.20E-05	6.76E-05	1.33E-04
σ_{v1}^2	2.00E-06	2.09E-06	9.77E-07	2.10E-06	3.09E-06
		5.94E-08	3.14E-08	1.43E-07	8.60E-07
RV1	2.00E-04	7.52E-02	4.76E-02	7.69E-03	4.52E-04
		1.14E-03	8.33E-04	2.90E-04	8.81E-05
σ_{x2}^2	2.00E-04	2.05E-04	2.06E-04	2.07E-04	2.10E-04
		4.26E-05	4.27E-05	6.70E-05	1.31E-04
σ_{v2}^2	2.00E-06	2.09E-06	9.78E-07	2.12E-06	3.09E-06
		5.75E-08	3.20E-08	1.51E-07	8.75E-07
RV2	2.00E-04	7.52E-02	4.76E-02	7.71E-03	4.52E-04
		1.13E-03	8.11E-04	2.99E-04	8.90E-05
σ_{x12}^2	1.00E-04	1.00E-04	9.99E-05	1.02E-04	1.07E-04
		3.60E-05	3.36E-05	5.11E-05	1.05E-04
σ_{v12}^2	0.00E+00	5.68E-10	−8.17E-10	1.48E-08	5.24E-07
		4.81E-08	1.98E-08	1.05E-07	6.01E-07
RCV	1.00E-04	8.59E-05	4.57E-05	1.02E-04	1.03E-04
		5.66E-04	3.99E-04	2.18E-04	6.46E-05
HY	1.00E-04	1.00E-04			
		4.03E-04			
RCV-RV	5.00E-01	1.14E-03	9.64E-04	1.32E-02	2.28E-01
		7.52E-03	8.40E-03	2.83E-02	1.37E-01
HY-RV	5.00E-01	1.34E-03	2.11E-03	1.31E-02	2.29E-01
		5.36E-03	8.46E-03	5.25E-02	9.48E-01
SIML-SIML	5.00E-01	4.91E-01	4.85E-01	4.99E-01	5.17E-01
		1.45E-01	1.30E-01	2.02E-01	5.21E-01

(**Step-1**) The first step is to argue that the effects of random sampling and drift terms in the underlying stochastic processes are stochastically negligible under our assumptions.

Although t_i^n and n^* are random variables, which are finite-valued and bounded in $[0, 1]$ for any n. We write $y_i^a = x_i^a + v_i^a$ with $a = s$ or f, where $y_i^a = y_a(t_i^n)$ and $x_i^a = X_a(t_i^n)$ in the basic case. We set $\mathbf{y}_i^n = (y^s(t_i^n), y^f(t_i^n))'$, $\mathbf{x}_i^n = (x^s(t_i^n), x^f(t_i^n))'$, and we write the underlying (unobservable) returns in the period $(t_{i-1}^n, t_i^n]$ as

$$\mathbf{r}_i^n = \mathbf{x}_i^n - \mathbf{x}_{i-1}^n = \int_{t_{i-1}^n}^{t_i^n} \boldsymbol{\mu}_x(s, X(s))ds + \int_{t_{i-1}^n}^{t_i^n} \mathbf{C}_x(s)d\mathbf{B}_s \quad (i = 1, \ldots, n^*) \quad (7.25)$$

and the martingale part as

Table 7.2 Estimation of covariance and hedging coefficient : Model 1 (ACD, $n \sim 18,000$)

18,000	True	Raw	1 s	10 s	300 s
σ_{x1}^2	2.00E-04	2.05E-04	2.06E-04	2.05E-04	2.03E-04
		4.30E-05	4.27E-05	6.64E-05	1.27E-04
σ_{v1}^2	2.00E-06	2.00E-06	9.32E-07	2.03E-06	2.98E-06
		5.55E-08	3.36E-08	1.41E-07	8.14E-07
RV1	2.00E-04	7.22E-02	4.51E-02	7.40E-03	4.38E-04
		1.62E-03	9.34E-04	3.01E-04	8.63E-05
σ_{x2}^2	2.00E-04	2.05E-04	2.06E-04	2.07E-04	2.12E-04
		4.25E-05	4.27E-05	6.85E-05	1.42E-04
σ_{v2}^2	2.00E-06	2.01E-06	9.31E-07	2.03E-06	3.04E-06
		5.58E-08	3.34E-08	1.46E-07	8.41E-07
RV2	2.00E-04	7.22E-02	4.52E-02	7.41E-03	4.45E-04
		1.78E-03	9.57E-04	3.02E-04	8.89E-05
σ_{x12}^2	1.00E-04	1.01E-04	1.00E-04	1.02E-04	1.03E-04
		3.70E-05	3.37E-05	5.13E-05	1.03E-04
σ_{v12}^2	0.00E+00	-9.58E-10	-3.15E-10	6.07E-09	4.83E-07
		4.56E-08	1.83E-08	1.03E-07	5.84E-07
RCV	1.00E-04	7.51E-05	3.09E-05	8.39E-05	1.00E-04
		5.01E-04	3.78E-04	2.16E-04	6.06E-05
HY	1.00E-04	1.31E-04			
		3.80E-04			
RCV-RV	5.00E-01	1.04E-03	6.88E-04	1.13E-02	2.32E-01
		6.94E-03	8.36E-03	2.91E-02	1.39E-01
HY-RV	5.00E-01	1.81E-03	2.90E-03	1.78E-02	3.10E-01
		5.26E-03	8.41E-03	5.14E-02	9.22E-01
SIML-SIML	5.00E-01	4.90E-01	4.85E-01	4.98E-01	5.24E-01
		1.46E-01	1.28E-01	2.06E-01	5.14E-01

$$\mathbf{r}_i^* = \int_{t_{i-1}^n}^{t_i^n} \mathbf{C}_x(s) d\mathbf{B}_s \quad (i = 1, \ldots, n^*) \tag{7.26}$$

with $0 = t_0 \le t_1^n < \cdots < t_{n^*}^n \le 1$. By using Assumptions I and II, we have $n^*/n = 1 + o_p(1)$,

$$\mathbf{E}[\|\mathbf{r}_j^n\|^2] = \mathbf{E}[\|\int_{t_{i-1}^n}^{t_i^n} \boldsymbol{\mu}_x(s) ds|^2] + 2\mathbf{E}\left[\left(\int_{t_{i-1}^n}^{t_i^n} \boldsymbol{\mu}_x(s) ds\right)' \mathbf{r}_i^*\right] + \mathbf{E}[\|\mathbf{r}_i^*\|^2],$$

and

$$\mathbf{E}[\|\int_{t_{i-1}^n}^{t_i^n} \mathbf{C}_x(s) d\mathbf{B}_s - \int_{t_{i-1}^n}^{t_i^n} \mathbf{C}_x(t_{i-1}^n) d\mathbf{B}_s\|^2] = O\left(\left(\frac{1}{n}\right)^2\right).$$

Table 7.3 Estimation of covariance and hedging coefficient : Model 3 ($b = 0.2, n \sim 18,000$)

18,000	True	Raw	1 s	10 s	300 s
σ_{x1}^2	2.00E-04	2.25E-04	2.28E-04	2.09E-04	2.15E-04
		4.53E-05	4.53E-05	6.83E-05	1.38E-04
σ_{v1}^2	2.00E-06	6.27E-07	5.63E-07	4.31E-06	6.53E-06
		1.78E-08	1.67E-08	3.12E-07	1.80E-06
RV1	2.00E-04	4.00E-02	3.63E-02	1.74E-02	8.59E-04
		5.26E-04	5.97E-04	7.05E-04	1.81E-04
σ_{x2}^2	2.00E-04	2.27E-04	2.29E-04	2.10E-04	2.13E-04
		4.74E-05	4.78E-05	6.83E-05	1.34E-04
σ_{v2}^2	2.00E-06	6.27E-07	5.63E-07	4.30E-06	6.45E-06
		1.73E-08	1.71E-08	2.99E-07	1.81E-06
RV2	2.00E-04	4.00E-02	3.63E-02	1.74E-02	8.55E-04
		5.37E-04	5.98E-04	6.80E-04	1.74E-04
σ_{x12}^2	1.00E-04	9.98E-05	9.96E-05	1.02E-04	1.06E-04
		3.80E-05	3.66E-05	5.16E-05	1.07E-04
σ_{v12}^2	0.00E+00	5.23E-11	-3.47E-10	5.64E-09	5.67E-07
		2.06E-08	1.14E-08	2.12E-07	1.30E-06
RCV	1.00E-04	2.40E-05	1.16E-05	6.91E-05	1.05E-04
		3.42E-04	2.87E-04	4.83E-04	1.30E-04
HY	1.00E-04	2.74E-05			
		3.75E-04			
RCV-RV	5.00E-01	5.94E-04	3.18E-04	3.96E-03	1.23E-01
		8.55E-03	7.91E-03	2.77E-02	1.50E-01
HY-RV	5.00E-01	6.83E-04	7.46E-04	1.57E-03	2.98E-02
		9.39E-03	1.04E-02	2.16E-02	4.63E-01
SIML-SIML	5.00E-01	4.43E-01	4.36E-01	4.89E-01	5.08E-01
		1.45E-01	1.36E-01	2.05E-01	4.93E-01

Then we can evaluate as

$$\mathbf{E}[\|\mathbf{r}_i^n\|^2 - \int_{t_{i-1}^n}^{t_i^n} \operatorname{tr}(\mathbf{\Sigma}_x(s))ds] = O\left(\left(\frac{1}{n}\right)^{3/2}\right). \tag{7.27}$$

Hence, we find that drift terms have negligible effects on the estimation of the integrated volatility function under Assumptions I and II.

Similarly, we can use that $n^*/n \overset{p}{\to} 1$ and

$$\sum_{j=1}^{n^*} \int_{t_{j-1}^n}^{t_j^n} \mathbf{\Sigma}_x(s)ds - \int_0^1 \mathbf{\Sigma}_x(s)ds \overset{p}{\to} O, \tag{7.28}$$

Table 7.4 Estimation of covariance and hedging coefficient : Model 4 ($\eta=0.001$, $n \sim 18,000$)

18,000	True	Raw	1 s	10 s	300 s
σ_{x1}^2	2.00E-04	2.04E-04	2.05E-04	2.06E-04	2.11E-04
		4.20E-05	4.19E-05	6.76E-05	1.34E-04
σ_{v1}^2	2.00E-06	2.09E-06	9.76E-07	2.11E-06	3.09E-06
		5.82E-08	3.12E-08	1.45E-07	8.68E-07
RV1	2.00E-04	7.52E-02	4.76E-02	7.69E-03	4.51E-04
		1.12E-03	8.45E-04	2.90E-04	8.90E-05
σ_{x2}^2	2.00E-04	2.05E-04	2.06E-04	2.06E-04	2.10E-04
		4.26E-05	4.26E-05	6.68E-05	1.31E-04
σ_{v2}^2	2.00E-06	2.09E-06	9.79E-07	2.12E-06	3.07E-06
		5.75E-08	3.22E-08	1.56E-07	8.75E-07
RV2	2.00E-04	7.52E-02	4.76E-02	7.71E-03	4.51E-04
		1.13E-03	8.21E-04	3.05E-04	8.98E-05
σ_{x12}^2	1.00E-04	1.00E-04	9.99E-05	1.02E-04	1.07E-04
		3.59E-05	3.35E-05	5.11E-05	1.05E-04
σ_{v12}^2	0.00E+00	-3.00E-10	-1.33E-09	1.39E-08	5.15E-07
		4.74E-08	1.97E-08	1.04E-07	6.03E-07
RCV	1.00E-04	7.50E-05	4.12E-05	1.02E-04	1.02E-04
		5.60E-04	3.98E-04	2.17E-04	6.47E-05
HY	1.00E-04	9.48E-05			
		3.99E-04			
RCV-RV	5.00E-01	9.95E-04	8.70E-04	1.32E-02	2.27E-01
		7.45E-03	8.37E-03	2.82E-02	1.38E-01
HY-RV	5.00E-01	1.26E-03	2.00E-03	1.23E-02	2.12E-01
		5.30E-03	8.38E-03	5.19E-02	9.31E-01
SIML-SIML	5.00E-01	4.91E-01	4.86E-01	4.99E-01	5.19E-01
		1.44E-01	1.30E-01	2.02E-01	5.17E-01

where upon

$$\mathrm{E}[\int_{t_{i-1}^n}^{t_i^n} \mathbf{\Sigma}_x(s)ds] = O\left(\frac{1}{n}\right). \tag{7.29}$$

We note that the volatility function $\mathbf{\Sigma}_x(s)$ ($0 \le s \le 1$) and the integrated volatility $\mathbf{\Sigma}_x = \int_0^1 \mathbf{\Sigma}_x(s)ds$ can be stochastic.

By above arguments, we advance the present proof as if n^* were fixed and were replaced by the corresponding fixed n.

Then we use the fixed n (and m_{n^*}) as if it were n^* (and m_n) in the following developments of this section for the sake of simplicity. (Basically we need to replace n by n^* in each step, and then we invoke many tedious arguments because n^* is stochastic.)

Table 7.5 Estimation of covariance and hedging coefficient : Model 5 ($b_1 = 0.2, b_2 = 5; n \sim$ 18, 000)

18,000	True	Raw	1 s	10 s	300 s
σ_{x1}^2	2.00E-04	2.15E-04	2.17E-04	2.08E-04	2.15E-04
		4.36E-05	4.35E-05	6.83E-05	1.36E-04
σ_{v1}^2	2.00E-06	7.24E-07	6.00E-07	3.59E-06	4.98E-06
		2.07E-08	1.81E-08	2.59E-07	1.37E-06
RV1	2.00E-04	4.29E-02	3.69E-02	1.37E-02	6.75E-04
		5.75E-04	6.15E-04	5.55E-04	1.40E-04
σ_{x2}^2	2.00E-04	2.16E-04	2.18E-04	2.09E-04	2.13E-04
		4.51E-05	4.53E-05	6.79E-05	1.33E-04
σ_{v2}^2	2.00E-06	7.23E-07	6.01E-07	3.59E-06	4.94E-06
		2.00E-08	1.86E-08	2.47E-07	1.39E-06
RV2	2.00E-04	4.29E-02	3.69E-02	1.37E-02	6.74E-04
		5.91E-04	6.10E-04	5.44E-04	1.37E-04
σ_{x12}^2	1.00E-04	1.01E-04	1.01E-04	1.03E-04	1.08E-04
		3.71E-05	3.51E-05	5.17E-05	1.07E-04
σ_{v12}^2	0.00E+00	6.82E-11	-4.05E-10	9.73E-09	5.53E-07
		2.26E-08	1.22E-08	1.81E-07	9.86E-07
RCV	1.00E-04	3.27E-05	1.98E-05	8.14E-05	1.05E-04
		3.54E-04	2.95E-04	3.88E-04	1.01E-04
HY	1.00E-04	3.81E-05	0.00E+00	0.00E+00	0.00E+00
		3.70E-04	0.00E+00	0.00E+00	0.00E+00
RCV-RV	5.00E-01	7.58E-04	5.39E-04	5.92E-03	1.56E-01
		8.26E-03	8.00E-03	2.84E-02	1.48E-01
HY-RV	5.00E-01	8.89E-04	1.03E-03	2.80E-03	5.56E-02
		8.63E-03	1.01E-02	2.71E-02	5.80E-01
SIML-SIML	5.00E-01	4.69E-01	4.63E-01	4.96E-01	5.14E-01
		1.44E-01	1.33E-01	2.04E-01	4.95E-01

(**Step-2**) Let $Z_{in}^{(1)}$ and $Z_{in}^{(2)}$ ($i = 1, \ldots, n$) be the ith elements of $\overset{*}{n} \times 2$ vectors

$$\mathbf{Z}_n^{(1)} = h_n^{-1/2} \mathbf{P}_n \mathbf{C}_n^{-1} (\mathbf{X}_n - \bar{\mathbf{Y}}_0) \, , \quad \mathbf{Z}_n^{(2)} = h_n^{-1/2} \mathbf{P}_n \mathbf{C}_n^{-1} \mathbf{V}_n, \qquad (7.30)$$

respectively, where $\mathbf{X}_n = (\mathbf{x}_i') = (x_i^s, x_i^f)$, $\mathbf{V}_n = (\mathbf{v}_i') = (v_i^s, v_i^f)$, $\mathbf{X}_n = (\mathbf{z}_{in}')$ are $n \times 2$ vectors with $\mathbf{z}_{in} = \mathbf{z}_{in}^{(1)} + \mathbf{z}_{in}^{(2)}$ and $\mathbf{P}_n$ is defined by (3.11).

We write $z_{kn}^{s(j)}$ and $z_{kn}^{f(j)}$ ($j = 1, 2$) as the first and second components of $\mathbf{z}_{kn}$, and we use the notations $z_{kn}^{s,j}$ and $z_{kn}^{f,j}$ ($j = 1, 2$) for the jth components of $\mathbf{z}_{kn}^s$ and $\mathbf{z}_{kn}^f$, respectively. Then we can use the arguments in Chap. 5.

(**Step-3**) We take $\hat{\sigma}_{ss}^{(x)}$ as the estimator of $\sigma_{ss}^{(x)}$ because we can use the similar arguments to $\sigma_{ff}^{(x)}$ and $\sigma_{sf}^{(x)}$. We give the asymptotic variance of the leading term of

$$\sqrt{m_n}\left[\hat{\sigma}_{ss}^{(x)} - \sigma_{ss}^{(x)}\right] = \sqrt{m_n}\left[(1/m_n)\sum_{k=1}^{m_n}(z_{kn}^s)^2 - \sigma_{ss}^{(x)}\right],$$

that is,

$$\sqrt{m_n}\left[\frac{1}{m_n}\sum_{k=1}^{m_n}(z_{kn}^{s,(1)})^2 - \sigma_{ss}^{(x)}\right] \tag{7.31}$$

because it is of order $O_p(1)$. We write

$$z_{kn}^{s,(1)} = \sqrt{\frac{n}{2n+1}}\sum_{j=1}^{n}r_j^s(e^{i\theta_{kj}} + e^{-i\theta_{kj}}), \tag{7.32}$$

where r_j^s is the first component of $\mathbf{r}_j^*$ and $\theta_{kj} = [2\pi/(2n+1)](k-1/2)(j-1/2)$. By using the relation that $(e^{i\theta_{kj}} + e^{-i\theta_{kj}})^2 = 2 + e^{2i\theta_{kj}} + e^{-2i\theta_{kj}}$, we represent

$$\left[\frac{2n+1}{2n}\right]\left[\frac{1}{m_n}\sum_{k=1}^{m_n}(z_{kn}^{s,(1)})^2 - \int_0^1\sigma_{ss}^{(x)}(s)ds\right] \tag{7.33}$$

$$= \frac{1}{m_n}\sum_{k=1}^{m_n}\left\{\frac{1}{2}\left[\sum_{j=1}^{n}(r_j^s)^2(e^{i\theta_{kj}} + e^{-i\theta_{kj}})^2 - 2\left(1+\frac{1}{2n}\right)\int_0^1\sigma_{ss}^{(x)}(s)ds\right.\right.$$

$$\left.\left.+\left[\sum_{j\neq j'=1}^{n}r_j^s r_{j'}^s(e^{i\theta_{kj}} + e^{-i\theta_{kj}})(e^{i\theta_{kj'}} + e^{-i\theta_{kj'}})\right]\right\}\right.$$

$$= 2\sum_{j>j'=1}^{n}r_j^s r_{j'}^s\left[\frac{1}{m_n}\sum_{k=1}^{m_n}(e^{i\theta_{kj}} + e^{-i\theta_{kj}})(e^{i\theta_{kj'}} + e^{-i\theta_{kj'}})\right]$$

$$+\sum_{j=1}^{n}\left[(r_j^s)^2 - \int_{t_{j-1}^n}^{t_j^n}\sigma_{ss}^{(x)}(s)ds\right]$$

$$+\frac{1}{2}\sum_{j=1}^{n}\left[(r_j^s)^2 - \int_{t_{j-1}}^{t_j^n}\sigma_{ss}^{(x)}(s)ds\right]\left[\sum_{k=1}^{m_n}(e^{2i\theta_{kj}} + e^{-2i\theta_{kj}})\right]$$

$$-\frac{1}{2n}\sum_{j=1}^{n}\left[\int_{t_{j-1}}^{t_j^n}\sigma_{ss}^{(x)}(s)ds\right]$$

$$= (A) + (B) + (C) + (D),\text{ (say)}.$$

Then by using the same derivations of Chap. 5, it is possible to show that except for the first term (A), we have $\sqrt{m_n}(B) \xrightarrow{P} 0$, $\sqrt{m_n}(C) \xrightarrow{P} 0$, and $\sqrt{m_n}(D) \xrightarrow{P} 0$ as $m_n \to \infty$ ($n \to \infty$). Also by using the simple relation

$$(e^{i\theta_{kj}} + e^{-i\theta_{kj}})(e^{i\theta_{kj'}} + e^{-i\theta_{kj'}})$$
$$= (e^{i(\theta_{kj}+\theta_{kj'})} + e^{-i(\theta_{kj}+\theta_{kj'})}) + (e^{i(\theta_{kj}-\theta_{kj'})} + e^{-i(\theta_{kj}-\theta_{kj'})}) ,$$

the random quantity $\sqrt{m_n}\left[\frac{1}{m_n}\sum_{k=1}^{m_n}(z_{kn}^{s,(1)})^2 - \sigma_{ss}^{(x)}\right]$ is asymptotically equivalent to

$$(A)' = 2\sum_{j>j'=1}^{n} r_j^s r_{j'}^s h_m(j, j') , \tag{7.34}$$

where for $j, j' = 1, \ldots, n$

$$h_m(j, j') = \frac{1}{2\sqrt{m_n}}\frac{\sin 2m(\theta_j + \theta_{j'})/2}{\sin(\theta_j + \theta_{j'})/2} + \frac{1}{2\sqrt{m_n}}\frac{\sin 2m(\theta_j - \theta_{j'})/2}{\sin(\theta_j - \theta_{j'})/2}$$

and $\theta_j = [2\pi/(2n + 1)](j - 1/2)$.

By extending the evaluation method on Fejér kernel (see Chap. 8 of Anderson (1971)), we can derive the variance of the asymptotic distribution of $(A)'$, which is asymptotically equivalent to (A).

Lemma 7.1 *Under Assumption II, as $n \to \infty$ the asymptotic variance of $(A)'$ is given by*

$$V_{ss} = 2\int_0^1 \left[\sigma_{ss}^{(x)}(\tau(s))\right]^2 d(s)ds . \tag{7.35}$$

Proof of Lemma 7.1 *From the representation of $(A)'$, given the sampling process t_j^n $(j = 1, \ldots, n^*)$ we have the conditional expectation*

$$E[((A)')^2|\{t_j^s\}]$$
$$= 2\sum_{j>j'=1}^{n^*} \sigma_{ss.j}^{(x)}\sigma_{ss.j'}^{(x)}\left[\frac{1}{4m}\right]\left[\frac{\sin 2m\pi(j + j' - 1)/(2^{n^*} + 1)}{\sin \pi(j + j')/(2^{n^*} + 1)}\right.$$
$$\left.+ \frac{\sin 2m\pi(j - j' - 1)/(2^{n^*} + 1)}{\sin \pi(j - j')/(2^{n^*} + 1)}\right]^2 ,$$

where we use the notation

$$\sigma_{ss.j}^{(x)} = E\left[\int_{t_{j-1}^n}^{t_j^n} \sigma_{ss}^{(x)}(s)ds\right] . \tag{7.36}$$

Under Assumption I by using the standard approximation argument for integration, we find that

$$E\left[\int_{t_{j-1}^n}^{t_j^n} \sigma_{ss}^{(x)}(s)ds - \int_{t_{j-1}^n}^{t_j^n} \sigma_{ss}^{(x)}(t_{j-1}^n)ds\right] = o\left(\frac{1}{n}\right) \tag{7.37}$$

and

$$\int_{t_{j-1}^n}^{t_j^n} \sigma_{ss}^{(x)}(t_{j-1}) ds = (t_n^n - t_{j-1}^n) \sigma_{ss}^{(x)}(t_{j-1}^n) = O_p\left(\frac{1}{n}\right). \tag{7.38}$$

For any bounded continuous functions $a(s^*)$ and $b(t^*)$ in $[0, 1]$, we utilized the representation

$$2\int_0^1 \int_0^1 \frac{1}{4m}\left[\frac{\sin m\pi(s^* - t^*)}{\sin(\pi/2)(s^* - t^*)} + \frac{\sin m\pi(s^* + t^*)}{\sin(\pi/2)(s^* + t^*)}\right]^2 a(s^*)b(^*t)ds^*dt^*$$

$$= 2\int_0^1 \int_0^1 \frac{1}{4m}\left\{\left[\frac{\sin m\pi(s^* - t^*)}{\sin(\pi/2)(s^* - t^*)}\right]^2 + \left[\frac{\sin m\pi(s^* + t^*)}{\sin(\pi/2)(s^* + t^*)}\right]^2\right.$$

$$\left. + \left[\frac{\sin m\pi(s^* - t^*)}{\sin(\pi/2)(s^* - t^*)}\right]\left[\frac{\sin m\pi(s^* + t^*)}{\sin(\pi/2)(s^* + t^*)}\right]\right\} a(s^*)b(t^*)ds^*dt^*$$

$$= (E) + (F) + (G) , \text{(say)}.$$

(See the proof of Lemma 5.6 in Chap. 5.) By changing the order of integration, as $m \to \infty$ we can evaluate the first term as

$$2\int_0^1 \frac{1}{2m}\left[\frac{\sin^2(2m)\frac{\pi}{2}u}{\sin^2(\pi/2)u}\right]\left[\int_0^{1-u} a(u + t^*)b(t^*)dt\right]du \longrightarrow 2\lim_{u\to 0}\int_0^{1-u} a(u + t^*)b(t^*)dt^*$$

and the second term is negligible because

$$(F) = 2\int_0^1 \frac{1}{2m}\left[\frac{\sin^2(2m)\frac{\pi}{2}u}{\sin^2(\pi/2)u}\right]\left[\int_0^u a(s^*)b(u - s^*)ds^*\right]du \xrightarrow{P} 0$$

as $m \to \infty$. By applying a similar argument to the third term (G), we find that it is also negligible when m is large. (These arguments are essentially the same as the ones in the proof of Lemma 5.6 in Chap. 5.)

We need to show the effects of discretization and random sampling because t_i^n are sequence of random variables. Under Assumptions I and II, as $(i(n) - 1)/n \to s$ and $(j(n) - 1)/n \to t$ for n being large while m is fixed, we have $t_i^n \xrightarrow{P} \tau(s)$, $t_j^n \to \tau(t)$, $n(t_i^n - t_{i-1}^n) \xrightarrow{P} d(s)$, and $n(t_j^n - t_{j-1}^n) \xrightarrow{P} d(t)$. Then the only non-negligible term corresponds to

$$V' = \int_0^1 \int_0^1 \frac{1}{4m}\left[\frac{\sin m\pi(s - t)}{\sin(\pi/2)(s - t)}\right]^2 \sigma_{ss}^{(x)}(\tau(s))\sigma_{ss}^{(x)}(\tau(t))d(s)d(t)dsdt . \tag{7.39}$$

Then by letting $m \to \infty$, we have the desired result. □

(Step-4) For this step, we need to show the stable convergence in law of (7.34), but the arguments are quite similar to those in the proof of Theorem 3.4, which is based on the method explained in Chap. 5. Hence, we have omitted the details of our arguments in this version.

(**Step-5**) Finally, we need to deal with the integrated covariance. By modifying the derivations for the proof of integrated covariance, we use $\hat{\sigma}_{ss}^{(x)}$, $\hat{\sigma}_{ff}^{(x)}$ and $\hat{\sigma}_{sf}^{(x)}$. (It is straightforward to develop arguments which are essentially the same for $\hat{\sigma}_{ss}^{(x)}$, and we have omitted the details.) Then we can evaluate the variance of the asymptotic distributions of integrated covariance SIML estimator and the resulting variance formula becomes

$$V_{sf} = \int_0^1 \left[\sigma_{ss}^{(x)}(\tau(s))\sigma_{ff}^{(x)}(\tau(s)) + (\sigma_{sf}^{(x)}(\tau(s)))^2 \right] d(s)ds . \qquad (7.40)$$

□

References

Anderson, T.W. 1971. *The statistical analysis of time series*. Berlin: Wiley.

Barndorff-Nielsen, O., P. Hansen, A. Lunde, and N. Shephard. 2011. Multivariate realized kernels: consistentpositive semi-definite estimators of the covariation of equity prices with noise and non-synchronous trading. *Journal of Econometrics* 162: 149–169.

Duffie, D. 1989. *Futures markets*. Prentice Hall.

Engle, R.F., and J.R. Russell. 1998. Autoregressive conditional duration models: a new model for irregularly spaced transaction data. *Econometrica* 66–5: 1127–1162.

Harris, F., T. McInish, G. Shoesmith, and R. Wood. 1995. Cointegration, error correction and price discovery on informationally-linked security markets. *Journal of Financial and Quantitative Analysis* 30: 563–581.

Hayashi, T., and N. Yoshida. 2005. On covariance estimation of non-synchronous observed diffusion processes. *Bernoulli* 11 (2): 359–379.

Chapter 8
Local SIML Estimation of Brownian Functionals

Abstract We introduce the local SIML (LSIML) method for estimating some Brownian functionals including the asymptotic variance of the SIML estimator. It is an extension of the basic SIML method and we show the usefulness of the LSIML method through simulations.

8.1 Introduction

In the previous chapters, we developed the SIML method of estimating the volatility and co-volatilities of security prices when the underlying processes are the class of diffusion processes. In this chapter, we extend the SIML method by developing the local SIML (LSIML) estimation, which is a new statistical method. The method of the LSIML estimation is to use blocking original data and averaging the SIML estimates of each block. The main motivation for developing this method is to estimate some Brownian functionals which are more general than the cases of the volatility and co-volatility. They are appeared in the asymptotic distributions of the SIML estimator, for instance. By using the local SIML method, we can estimate the asymptotic variance of the SIML estimator. More generally, by using the LSIML method, it is possible to estimate higher-order Brownian functionals which play an important role in high-frequency econometric problems.

The LSIML method has reasonable finite-sample properties, which are illustrated by several simulations. Because the LSIML estimation is a straightforward extension of the SIML estimation, it is quite simple and so is useful in practical applications. Besides, the LSIML method may have desirable asymptotic properties such as consistency and asymptotic normality under a set of additional assumptions. In this chapter, we give only some results of simulations on the LSIML estimation.

© The Author(s) 2018 97
N. Kunitomo et al., *Separating Information Maximum Likelihood Method for High-Frequency Financial Data*, JSS Research Series in Statistics, https://doi.org/10.1007/978-4-431-55930-6_8

8.2 Estimation of Brownian Functionals

In this chapter, we consider the basic case and the deterministic time-varying case when $p = q = 1$. Let

$$Y(t_i^n) = X(t_i^n) + v(t_i^n) \quad (i = 1, \ldots, n) \tag{8.1}$$

be the (one dimensional) observed (log-)price at t_i^n $(0 = t_0^n \le t_1^n \le \cdots \le t_n^n = 1)$ and $v(t_i^n)$ $(= v_i)$ be a sequence of i.i.d. random variables with $\mathbf{E}[v_i] = 0$ and $\mathbf{E}[v_i^2] = \sigma_v^2$ (> 0).

The underlying continuous-time Brownian martingale is

$$X(t) = X(0) + \int_0^t \sigma_s dB_s \quad (0 \le s \le t \le 1) , \tag{8.2}$$

which is independent of $v(t_i^n)$, and σ_s is the (instantaneous) volatility function, which can be deterministic time-varying cases. Although it may be possible to apply the LSIML method to more general Itô semi-martingales, we will consider only this situation because it gives the essential feature of the LSIML method in a simple way.

The problem of our original interest is how to estimate Brownian functionals of the form

$$V(2r) = \int_0^1 \sigma_s^{2r} ds \tag{8.3}$$

for any positive integer r from a set of observations of $Y(t_i^n)$ $(i = 1, \ldots, n)$. There are important examples of this type of Brownian functional. An obvious example is the integrated volatility that corresponds to the case when $r = 1$.

Example 8.1 When r=1, we have the integrated volatility and it is given by

$$V(2) = \int_0^1 \sigma_s^2 ds . \tag{8.4}$$

Example 8.2 The asymptotic variance of the SIML estimator of integrated volatility $V(2)$ is given by

$$2V(4) = 2 \int_0^1 \sigma_s^4 ds . \tag{8.5}$$

It should be noted that estimating $V(4)$ with $r = 2$ is non-trivial task for which the SIML estimation cannot be used. Ait-Sahalia et al. (2005), and Ait-Sahalia and Jacod (2014) discussed some estimation methods of higher-order Brownian functionals, but it seems that they are more complicated than the method developed herein.

8.3 Local SIML Estimation

We extend the standard SIML method developed in Chap. 3. For simplicity, we take $t_j^n - t_{j-1}^n = 1/n$ $(j = 1, \ldots, n)$ and $t_0^n = 0$. We divide $(0, 1]$ into $b(n)$ sub-intervals, and in every interval, we allocate $c(n)$ observations. First, we consider the sequence $c^*(n)$ such that $c^*(n) \to \infty$ and we can take $b(n) \to \infty$ and $b(n) \sim n/c^*(n)$. A typical choice of observations in each interval would be $c^*(n) = [n^\gamma]$ $(0 < \gamma < 1)$, whereupon $b(n) \sim n^{1-\gamma}$. Because there are some extra observations (n is not equal to $b(n)c^*(n)$) and $b(n)$ is a positive integer, we adjust the number of terms in each interval $c(n) = c^*(n) + \text{(several terms)}$. We can ignore the effects of extra terms because they are asymptotically negligible and $b(n)c(n) = n$.

By setting $m_c = [c(n)^\alpha]$ $(0 < \alpha < 0.5)$, in the ith interval $(i = 1, \ldots, b(n))$ we apply the SIML transformation such that the transformed data are denoted as $z_k(i)$ $(k = 1, \ldots, c(n); i = 1, \ldots, b(n))$. Let the $2r$th moment in the ith interval is

$$M_{2r,n}(i) = \frac{1}{m_c} \sum_{k=1}^{m_c} [z_k(i)]^{2r} . \tag{8.6}$$

Then, we define the LSIML estimator of $V(2r)$ by

$$\hat{V}(2r) = \frac{b(n)^{r-1}}{a_r} \sum_{i=1}^{b(n)} M_{2r,n}(i) \tag{8.7}$$

where

$$a_r = \frac{2r!}{r! \, 2^r} . \tag{8.8}$$

In particular, $a_1 = 1$, $a_2 = 3$, and $a_3 = 15$.

If we take $c(n) = n$, $b(n) = 1$, and $r = 1$, then we have the SIML estimator for integrated volatility.

In this construction of the LSIML estimator, we need to normalize (8.6) due to the scale factor $1/n$ and the local Gaussianity of underlying continuous martingales.

For the LSIML estimator, we may expect the result that $\hat{V}(2r) \xrightarrow{p} V(2r)$ as $n \to \infty$. Furthermore, because in the special cases when $r = 1$ the SIML estimator have desirable asymptotic properties, we expect that we have the asymptotic normality as well as the consistency of $\hat{V}(2r)$ for $r > 1$ with some appropriate condition on m_c.

There can be different ways to construct the localizing SIML estimation, but we omit the discussion on the details.

8.4 Simulation

As an experimental exercise, we have done some simulation when $r = 1$ and $r = 2$, for the true parameters $V(2)$ and $3V(4)$. We note that the variance of the SIML estimator of integrated volatility corresponds to $2\hat{V}(4)$. We set $b(n) = [n^{1-\gamma}]$, $c(n) = $

$[n^\gamma]$ $(\gamma = 2/3)$, $n = 10,000$, and the number of replications is 5000. Also we have investigated several cases in which the instantaneous volatility function $\sigma_x^2(s)$ is given by

$$\sigma_s^2 = \sigma(0)^2 \left[a_0 + a_1 s + a_2 s^2 \right], \tag{8.9}$$

where a_i $(i = 0, 1, 2)$ are constants and we have some restrictions such that $\sigma_x(s)^2 > 0$ for $s \in [0, 1]$. This is a typical time-varying (but deterministic) case and the integrated volatility σ_x^2 is given by

$$\sigma_x^2 = \int_0^1 \sigma_s^2 ds = \sigma_x(0)^2 \left[a_0 + \frac{a_1}{2} + \frac{a_2}{3} \right]. \tag{8.10}$$

In this example, we have taken several intra-day volatility patterns including the flat (or constant) volatility, the monotone (decreasing or increasing) movements and the U-shaped movements.

In the following tables, the true parameter values of M2, 3M4, and 3LM4 are $\int_0^1 \sigma_s^2 ds$, $3(\int_0^1 \sigma_s^2 ds)^2$, and $3 \int_0^1 \sigma_s^4 ds$, respectively. In Tables 8.1, 8.2, and 8.3, the values of 3M4 and 3LM4 are 2, 12, and 21.6, respectively, while in Table 8.4 they are 2, 12, and 12, respectively.

In Tables 8.1, 8.2, 8.3, and 8.4, we first confirm that the LSIML method work well for the estimation of the integrated volatility. Although there may be some loss of estimation accuracy, the LSIML method gives desirable finite and asymptotic properties. The most important result in our simulation is the estimation of 3LM4,

Table 8.1 Estimation of integrated fourth-order functional ($a_0 = 6$, $a_1 = -24$, $a_2 = 24$; $\sigma_u^2 = 0.00$)

n=10,000	$\sigma_x^2 = 2.00$	3M4	3LM4
mean	1.99779323	11.964321	21.517617
SD	0.20525937	6.528366	6.329717

Table 8.2 Estimation of integrated fourth-order functional ($a_0 = 6$, $a_1 = -24$, $a_2 = 24$; $\sigma_u^2 = 1.0E-03$)

n=10,000	$\sigma_x^2 = 2.00$	3M4	3LM4
mean	2.08893559	12.07831	22.529918
SD	0.20769355	6.487159	6.456395

Table 8.3 Estimation of integrated fourth-order functional ($a_0 = 6$, $a_1 = -24$, $a_2 = 24$; $\sigma_u^2 = 1.0E-04$)

n=10,000	$\sigma_x^2 = 2.00$	3M4	3LM4
mean	2.00795529	11.97925	21.556694
SD	0.20264466	6.48342	6.301122

Table 8.4 Estimation of integrated fourth-order functional ($a_0 = 2, a_1 = 0, a_2 = 0; \sigma_u^2 = $ 1.0E-04)

n=10,000	$\sigma_x^2 = 2.00$	3M4	3LM4
mean	2.00454578	11.925724	12.03774
SD	0.15207006	4.904944	2.10631

which is (3/2) times the asymptotic variance of the SIML estimator of integrated volatility. As we see in Tables, the mean and standard deviation (SD) have reasonable values. It is interesting to find that the variance of 3LM4 is smaller than the one of 3M4.

From our numerical experiments, it seems that we need more than 20 blocks to estimate the second-order Brownian functionals. The SIML estimation case, however, we do not need this kind of requirements.

In any case, from our simulations the LSIML estimator of integrated volatility σ_x^2 performs quite well as we expected. The behaviors of the LSIML estimator for higher Brownian functionals as $r = 2$ are reasonable given the difficulties of the problem involved. Although we did not discuss the details, there is an interesting estimation problem of higher-order Brownian functionals when we have the stochastic volatility cases.

References

Ait-Sahalia, Y., and J. Jacod. 2014. *High-frequency financial econometrics*. Princeton University Press.

Ait-Sahalia, Y., P. Mykland, and L. Zhang. 2005. How often to sample a continuous-time process in the presence of market microstructure noise. *The Review of Financial Studies* 18–2: 351–416.

Chapter 9
Estimating Quadratic Variation Under Jumps and Micro-market Noise

Abstract We consider the estimation of quadratic variation of Itô's semi-martingales with jumps, which is an extension of volatility estimation in previous chapters. The SIML estimation gives reasonable estimation results of quadratic variation with jumps since it has desirable asymptotic properties such as consistency and asymptotic normality.

9.1 Introduction

In the previous chapters, we developed the SIML estimation for estimating the volatility and co-volatilities of security prices when the underlying processes are the class of diffusion processes. In this chapter, we investigate some functionals including the quadratic variation and quadratic covariation when there can be jump terms. When we can have jump terms in the underlying stochastic processes, an important class of stochastic processes with continuous-time is the Itô's semi-martingale. The theory of Itô's semi-martingale process has been developed as the general theory of stochastic processes, see Ikeda and Watanabe (1989) or Jacod and Protter (2012). In the theory of Ito's semi-martingales, the quadratic variation of the underlying process, which can be regarded an extension of the realized volatility, plays important roles as the fundamental quantities. Thus, it is important to estimate the quadratic variation (QV) and covariations from the discrete set of observed time series data.

There are some discussions on testing the presence of jumps in the underlying processes. Kurisu (2017) has utilized the SIML method for investigating testing procedures for jumps when there can be micro-market noise.

© The Author(s) 2018
N. Kunitomo et al., *Separating Information Maximum Likelihood Method for High-Frequency Financial Data*, JSS Research Series in Statistics, https://doi.org/10.1007/978-4-431-55930-6_9

9.2   SIML Estimation of Quadratic Covariation

Let $\mathbf{Y}(t_i^n)$ be a $p \times 1$ vector of observed (log-)prices at t_i^n ($0=t_0^n \le t_1^n \le \cdots \le t_n^n = 1$), which satisfy

$$\mathbf{Y}(t_i^n) = \mathbf{X}(t_i^n) + \mathbf{v}(t_i^n) \quad (i = 1, \ldots, n), \tag{9.1}$$

where $\mathbf{v}(t_i^n)$ is a $p \times 1$ vector sequence of (mutually) independent micro-market noise with $\mathbf{E}[\mathbf{v}(t_i^n)] = \mathbf{0}$ and $\mathbf{E}[\mathbf{v}(t_i^n)\mathbf{v}'(t_i^n)] = \mathbf{\Sigma}_v (> 0)$.

The underlying hidden process $\mathbf{X}(t)$ is a p-dimensional vector process, and it is an Itô's semi-martingale

$$\mathbf{X}(t) = \mathbf{X}(0) + \int_0^t \boldsymbol{\mu}_s ds + \int_0^t \mathbf{C}_x(s) \, d\mathbf{B}(s) + \int_s \int_{\|\delta(s,x)\|<1} \delta(s,\mathbf{x})(\mu - v)(ds, d\mathbf{x})$$

$$+ \int_s \int_{\|\delta(s,x)\|\ge 1} \delta(s,\mathbf{x})\mu(ds, d\mathbf{x}) . \tag{9.2}$$

As we introduced at the end of Chap. 2, $\boldsymbol{\mu}_s$ ($p \times 1$) and $\mathbf{C}_x(s)$ ($p \times q$) are the drift vector and volatility matrix, respectively, and $\mathbf{B}(t)$ is a $q \times 1$ vector of Brownian motions, $\delta(s, x)$ is a $p \times 1$ predictable function vector, $\mu(\cdot)$ is a $p \times 1$ vector of jump measure, and $v(\cdot)$ is the $p \times 1$ vector of compensator of $1_A * \mu$ for $1 * v(\omega)_t = v(\omega : [0, t) \times A)$ as we have used the notation of Ikeda and Watanabe (1989), and Jacod and Protter (2012).

When $p = q = 1$, the fundamental quantity for the continuous-time Itô's semi-martingale is the quadratic variation (QV), which is given by

$$U_0 = \int_0^1 \sigma_s^2 ds + \sum_{0 \le s \le 1} (\Delta X_s)^2 , \tag{9.3}$$

where we denote $\sigma_s = C_x(s)$.

If there are no jump terms, QV is equivalent to the integrated volatility. When there can be jump terms, however, it has been known in stochastic analysis that QV plays a fundamental role in the class of Ito's semi-martingales.

For the general case with $p \ge 1$, we have developed the SIML estimation when there are no jump terms. When we have p-dimensional jump terms $\Delta \mathbf{X}(s)$ ($= (\Delta X_g(s))$),

$$\mathbf{U}_0 = \int_0^1 \mathbf{\Sigma}(s)ds + \sum_{0 \le s \le 1} \Delta \mathbf{X}(s)\Delta \mathbf{X}'(s) . \tag{9.4}$$

When there is no jump, the SIML estimator of $\hat{\mathbf{\Sigma}}_x$ for the general $p \times p$ integrated volatility matrix $\mathbf{\Sigma}(s) = \mathbf{C}_x(s)\mathbf{C}_x'(s)$ ($= (\sigma_{gh}(s))$) and

$$\mathbf{\Sigma}_x = \int_0^1 \mathbf{\Sigma}(s)ds \tag{9.5}$$

is defined by

$$\hat{\boldsymbol{\Sigma}}_x = \frac{1}{m_n} \sum_{k=1}^{m_n} \mathbf{z}_k \mathbf{z}_k' = (\hat{\sigma}_{gh}^{(x)}) , \qquad (9.6)$$

where $\mathbf{z}_k = (z_{gk})$ $(g = 1, \ldots, p; k = 1, \ldots, m_n)$, which are constructed by the transformation from $\mathbf{Y}_n = (\mathbf{y}_k')$ to $\mathbf{Z}_n$ $(= (\mathbf{z}_k'))$ as we have introduced in Chap. 3.

In the case when there is no jump term (i.e., the Brownian Ito semi-martingale), we have shown in Chap. 3 that the SIML estimator is consistent and asymptotically normal in the stable convergence sense. For $m_n = [n^\alpha]$ and $0 < \alpha < 0.4$, as $n \longrightarrow \infty$

$$\sqrt{m_n} \left[\hat{\sigma}_{gh}^{(x)} - \int_0^1 \sigma_{gh}(s)ds \right] \overset{d}{\to} N\left[0, V_{gh}\right] \qquad (9.7)$$

in the stable convergence sense, where

$$V_{gh} = \int_0^1 \left[\sigma_{gg}^{(x)}(s)\sigma_{hh}^{(x)}(s) + \sigma_{gh}^{(x)2}(s) \right] ds . \qquad (9.8)$$

When X is an Itô semi-martingale with jumps, the asymptotic properties of the SIML estimator have not been given. In this respect, we have the next proposition in this situation and the outline of derivation is in the next subsection.

Proposition 9.1 *Assume (9.1) and (9.2) with additional conditions as in Theorem 3.4 (i.e., (3.38) and the related conditions), and let* $\hat{\boldsymbol{\Sigma}}_x = (\hat{\sigma}_{gh}^{(x)})$ *be the SIML estimator given by (3.18).*

(i) For $m_n = [n^\alpha]$ *and* $0 < \alpha < 0.5$, *as* $n \longrightarrow \infty$

$$\hat{\boldsymbol{\Sigma}}_x - \left[\int_0^1 \Sigma(s)ds + \sum_{0<s\leq 1} (\Delta X(s))(\Delta X(s))' ds \right] \overset{p}{\longrightarrow} \mathbf{O} . \qquad (9.9)$$

(ii) For $m_n = [n^\alpha]$ *and* $0 < \alpha < 0.4$, *as* $n \longrightarrow \infty$

$$\sqrt{m_n} \left[\hat{\sigma}_{gh}^{(x)} - \left(\int_0^1 \sigma_{gh}(s)ds + \sum_{0<s\leq 1} \Delta X_g(s)\Delta X_h(s) \right) \right] \overset{d}{\to} N\left[0, V_{gh}\right] \qquad (9.10)$$

in the stable convergence sense, where

$$V_{gh} = \int_0^1 \left[\sigma_{gg}^{(x)}(s)\sigma_{hh}^{(x)}(s) + \sigma_{gh}^{(x)2}(s) \right] ds \qquad (9.11)$$

$$+ \sum_{0<s\leq 1} \left[\sigma_{gg}^{(x)}(s)(\Delta X_h(s))^2 + \sigma_{hh}^{(x)}(s)(\Delta X_g(s))^2 \right.$$

$$\left. + 2\sigma_h^{(x)}(s)\sigma_g^{(x)}(s)(\Delta X_g(s)\Delta X_h(s)) \right] .$$

Corollary 9.1 *When $p = 1$, the asymptotic variance V is given by*

$$V = 2 \left[\int_0^1 \sigma_s^4 ds + 2 \sum_{0 < s \le 1} \sigma_s^2 (\Delta X(s))^2 \right] . \tag{9.12}$$

9.3 An Outline of the Derivation of Proposition 9.1

We give an intuitive argument for the above results. The basic method of proof is essentially the same to the one based on the derivations and their extensions given in Chap. 5. In the following, we only discuss the essential role of return vector process $\mathbf{r}_i = \mathbf{X}(t_i^n) - \mathbf{X}(t_{i-1}^n)$ $(i = 1, \ldots, n)$ in the decomposition of $\mathbf{Z}_n^{(1)}$. We omit the effects of c_{ij} $(i, j = 1, \ldots, n)$, for instance.

First, we consider the case when there does not exist any micro-market noise. Let $p \times p$ matrices $\mathbf{A}_n = (A_n(gh))$ and $\mathbf{A} = (A(gh))$ be

$$A_n(gh) = \sum_{i=1}^n (X_{gi} - X_{g,i-1})(X_{hi} - X_{h,i-1}) \tag{9.13}$$

and

$$A(gh) = \int_0^1 \sigma_{gh}(s) ds + \sum_{0 < s \le 1} \Delta X_{gi} \Delta X_{hi} , \tag{9.14}$$

where we set $\mathbf{X}(0) = (X_g(0))$, $\mathbf{X}_i = (X_g(t_i^n))$, $\Delta \mathbf{X}_i = (\Delta X_g(t_i^n))$, and $\mathbf{\Sigma}(s) = (\sigma_{gh}(s))$.

Let also $q = 1$ be the number of Brownian motions for simplicity. By using the same arguments in Chap. 5, we approximate

$$\sqrt{n} \left[A_n(gh) - A(gh) \right]$$

$$\sim \sqrt{n} \left\{ \sum_{i=1}^n \left[c_{g1}^\dagger(t_{i-1}^n)(B_i - B_{i-1}) + \sum_{t_{i-1}^n < s \le t_i^n} \Delta X_g(s) \right] \right.$$

$$\left. \times \left[c_{h1}^\dagger(t_{i-1}^n)(B_i - B_{i-1}) + \sum_{t_{i-1}^n < s \le t_i^n} \Delta X_h(s) \right] \right\} - A(gh) ,$$

where $\mathbf{C}_x(s) = (c_{g1}^\dagger(s))$ and $B_i = B(t_i^n)$. Then, the above quantity can be decomposed into

$$\sqrt{n}\left[\sum_{i=1}^{n} c_{g1}^{\dagger}(t_{i-1}^{n})c_{h1}^{\dagger}(t_{i-1}^{n})(B_i - B_{i-1})^2 - \int_0^1 \sigma_{gh}(s)ds\right]$$

$$+\sqrt{n}\left[\sum_{i=1}^{n} c_{g1}^{\dagger}(t_{i-1}^{n})(B_i - B_{i-1}) \sum_{t_{i-1}^{n} < s \leq t_i^{n}} \Delta X_h(s)\right]$$

$$+\sqrt{n}\left[\sum_{i=1}^{n} c_{h1}^{\dagger}(t_{i-1}^{n})(B_i - B_{i-1}) \sum_{t_{i-1}^{n} < s \leq t_i^{n}} \Delta X_g(s)\right].$$

Then, we denote (I), (II), and (III) in each terms of the last equality, and we can evaluate the asymptotic distributions.

Next, we apply the method of evaluating the asymptotic distributions of the SIML estimator used in Chap. 5. Then, the variance of the limiting random variables can be calculated as the variance of the above three terms except the factor $\sqrt{m_n}$ instead of $\sqrt{n}$ as stated in Proposition 9.1. In fact, we need to evaluate the effects of c_{ij} ($i, j = 1, \ldots, n$) as in Chap. 5, which are omitted here. It is because the resulting calculations in the general case with $q \geq 1$ are quite tedious, but they are straightforward as we have given the details for the diffusion cases in Chap. 5.

9.4 Some Numerical Analysis

We have performed a set of simulations with the general Itô semi-martingale processes with jumps. For the simulation method of jumps in the continuous-time stochastic process framework, see Cont and Tankov (2004).

Let $X(t) = (X_t^{(1)}, X_t^{(2)})'$ be a two-dimensional stochastic process. In particular, we simulated a set of realizations of the two-dimensional Itô semi-martingale given by

$$\begin{aligned} dX_1(t) &= c_1(t)dB_1(t) + Z_1(t)dN_1(t) + Z_3(t)dN_3(t) \\ dX_2(t) &= c_2(t)dB_2(t) + Z_2(t)dN_2(t) + Z_4(t)dN_3(t), \end{aligned} \quad (9.15)$$

where $\mathbf{B} = (B_1, B_2)'$ is two-dimensional Brownian motion and N_j for $j = 1, 2, 3$ are Poisson processes with intensities λ_j that are mutually independent and also independent of $\mathbf{B}$.

We use $Z = (Z_1, Z_1, Z_3, Z_4)'$ as the vector of jump sizes, which are cross sectionally, temporally, and independently distributed with laws F_{Z_j}. In this simulation, we set $\lambda_1 = \lambda_2 = 10$ and $\lambda_3 = 15$, and the jump size distributions are $Z_1(t), Z_2(t) \sim N(0, 5^{-2})$, and $Z_3(t), Z_4(t) \sim N(0, 10^{-2})$ as Gaussian distributions.

For the volatility process $c_j(t)$, we set

$$d(c_j(t)^2) = \alpha_j(\beta_j - c_j(t)^2)dt + \kappa_j c_j(t)dB_j^{\sigma}(t) \quad (j = 1, 2), \quad (9.16)$$

where $\alpha_1 = 3$, $\alpha_2 = 4$, $\beta_1 = 0.8$, $\beta_2 = 0.7$, $\kappa_1 = \kappa_2 = 0.5$, $\mathbf{E}[dB_j^{\sigma}(t)dB_j(t)] = \rho_j dt$ $(j = 1, 2)$, $\rho_1 = -0.5$, and $\rho_2 = -0.4$. As the market micro-structure noise, we use independent Gaussian noises for each component, that is, $\mathbf{v}(t_i^n) = (v_1(t_i^n), v_2(t_i^n))' \sim N_2(\mathbf{0}, 10^{-6}\mathbf{I}_2)$, where $\mathbf{I}_2$ is the identity matrix.

Simulation Procedure

For $i = 1$ and 2, let

$$[X, X]_{(i,i)}(t) = \int_0^t c_{(i,i)}(s)ds + \sum_{0 \leq s \leq t} (\Delta X_i(s))^2$$

$$V_{(i,i)}(t) = 2\left[\int_0^t (c_{(i,i)}(s))^2 ds + 2\sum_{0 \leq s \leq t} (c_{(i,i)}(s))(\Delta X_{(i)}(s))^2\right],$$

$$= 2\int_0^t (c_{(i,i)}(s))^2 ds$$

$$+ 4\left[\sum_{p:1 \leq T_p^{(i)} \leq t} (c_{(i,i)}(T_p^{(i)})(\Delta X_i(T_p^{(i)}))^2 + \sum_{p:1 \leq T_p^{(12)} \leq t} (c_{(i,i)}(T_p^{(12)})(\Delta X_i(T_p^{(12)})^2\right],$$

where $\{T_p^{(i)}\}$ $(i = 1, 2)$ are jump times when only the ith component jumps, and $\{T_p^{(12)}\}$ are jump times when both first and second components jump.

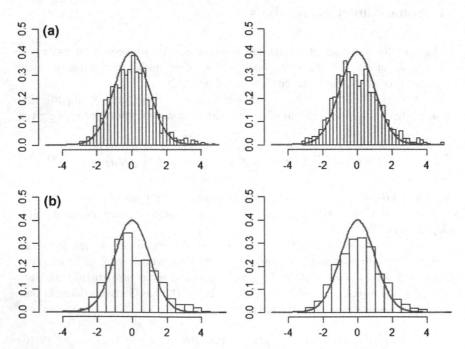

Fig. 9.1 **a** Finite sample distributions (Case 1). **b** Finite sample distributions (Case 2)

Let n be the sample size, and we plot the following values to check the validity of the central limit theorem (CLT) for the normalized random variables as

$$U_{1i} = (V_{(i,i)}(1)/n^{0.39})^{-1/2} \left(\widehat{[X, X]}^{(i,i)}_{1,SIML} - [X, X]^{(i,i)}_1 \right), \quad i = 1, 2, \qquad (9.17)$$

where $\widehat{[X, X]}^{(i,i)}_{1,SIML}$ is a (i, i) component of SIML estimator of the quadratic variation.

Simulation Results

We have plotted the histograms of the normalized estimator U_{11} and U_{12}. Although we performed number of simulations, we only have chosen two cases and in each figure the colored curve corresponds to the density of standard normal distribution. We give the simulation results for two cases, namely (a) Case 1 : $N = 1000$, $\Delta_n = 1/n$, $\mathbf{v}(t_i^n) \sim N_2(0, 10^{-6}\mathbf{I}_2)$ and (b) Case 2 : $N = 1000$, $\Delta_n = 1/n$, $\mathbf{v}(t_i^n) \sim N_2(0, 10^{-2}\mathbf{I}_2)$, where $n = 10,000$ and N is the number of simulations.

As we see in Fig. 9.1a, b in our simulations, we confirm that the SIML estimator has good finite sample as well as its asymptotic properties. The bias and the asymptotic variance of the SIML estimator agree with Proposition 9.1 of this chapter. Hence, we find that the simulation results are consistent with the asymptotic properties of the SIML estimator for the quadratic variation and covariation of the underlying stochastic process although we have jumps and micro-market noise term at the same time.

References

Cont, R., and P. Tankov. 2004. *Financial modeling with jump processes*. Chapman and Hall.

Ikeda, N., and S. Watanabe. 1989. *Stochastic differential equations and diffusion processes*, 2nd ed. North-Holland.

Jacod, J., and P. Protter. 2012. *Discretization of processes*. Berlin: Springer.

Kurisu, D. 2017. Power variations and testing for co-jumps: the small noise approach. *Scandinavian Journal of Statistics*. http://onlinelibrary.wiley.com/doi/10.1111/sjos.12309/abstract.

Chapter 10
Concluding Remarks

Abstract We conclude that the SIML estimation method we introduced in this book has good finite-sample as well as asymptotic properties. Because the SIML method is simple, it would be useful for practical applications. We mention further problems to be investigated.

Herein, we have developed a new statistical method for estimating integrated volatility and integrated covariances by using high-frequency financial data in the presence of noise. The SIML estimator proposed by Kunitomo and Sato (2008a, 2013), which was the origin of this book, can be regarded as a modification of the standard ML method and has many merits in the statistical sense. The SIML estimator has reasonable asymptotic properties; it is consistent and asymptotically normal (or has stable convergence in the general case) when the sample size is large and the data frequency interval is small under some conditions including non-Gaussian processes and volatility models. As we have demonstrated herein, the SIML estimator also has reasonable finite-sample properties and is asymptotically robust. We have omitted some details of the results, some of which are given by published papers, Kunitomo and Sato (2011, 2013), Misaki and Kunitomo (2015), Kunitomo et al. (2015), and unpublished reports, Kunitomo and Sato (2008a, b, 2010).

The SIML estimator is so simple that it can be used practically not only for the integrated volatility, but also for the covariations of multivariate high-frequency financial series. There may be potentially applications, and as an example, we showed a statistical analysis of high-frequency data of the Nikkei-225 Futures at OSE. We have confirmed that the presence of micro-market noises is an important factor in the market of Nikkei-225 Futures.

The original motivation to develop the SIML estimation method of volatility and co-volatilities was to measure financial risk in an appropriate way and to apply it for the risk managements and derivative pricing. When the volatility and co-volatility are stochastic and there are jumps in the underlying stochastic process, the standard finance theory based on the observed measure P and the equivalent (unique) martingale measure Q as we briefly explained in Chap. 2 is broken down. When we depart from the classical BS model and there are jumps, there can be an infinite number of equivalent martingale measures and then it is not clear how to develop the derivatives

© The Author(s) 2018
N. Kunitomo et al., *Separating Information Maximum Likelihood Method for High-Frequency Financial Data*, JSS Research Series in Statistics, https://doi.org/10.1007/978-4-431-55930-6_10

theory (see (Miyahara 2012) for instance). Although there has been a large literature on the related problem in mathematical finance, there is no consensus on the resulting procedure at this moment and thus there are interesting problems remained to utilize the statistical analysis of high-frequency financial data. It is certainly true that the volatility, co-volatility, and the quadratic variations are useful quantities for financial risk managements as we discussed on the use of hedging coefficient in Chap. 4.

As indicated in Chaps. 8 and 9, the SIML estimation method can apply to a number of interesting problems. Although the present discussions of these chapters are incomplete, there will be possible extensions and hence could be developed further. For instance, we did not analyze the problems around jumps in detail and determine the number of factors for covariation of high-frequency financial data. Also, we did not explain alternative methods proposed in the literature such as the realized kernel method, the pre-averaging method, for estimating volatility and covariance except brief comments in Chap. 6 which differ from the SIML method. The more systematic comparison with the SIML method may be an interesting topic although many of the existing methods are rather complicated in our view. Although there are interesting problems to be solved, further developments require new aspects and considerably more space, and we will discuss them on another occasion.

References

Kunitomo, N. and S. Sato 2008a. Separating information maximum likelihood estimation of realized volatility and covariance with micro-market noise, Discussion paper CIRJE-F-581, Graduate School of Economics, University of Tokyo. http://www.e.u-tokyo.ac.jp/cirje/research/dp/2008.

Kunitomo, N., and S. Sato. 2008b. *Realized Volatility, Covariance and Hedging Coefficient of Nikkei-225 Futures with Micro-Market Noise, Discussion Paper CIRJE-F-601*. Graduate school of economics: University of Tokyo.

Kunitomo, N. and S. Sato 2010. Robustness of the separating information maximum likelihood estimation of realized volatility with micro-market noise, CIRJE Discussion Paper F-733, University of Tokyo.

Kunitomo, N., and S. Sato. 2011. The SIML estimation of realized volatility of Nikkei-225 futures and hedging coefficient with micro-market noise. *Mathematics and Computers in Simulation* 81: 1272–1289.

Kunitomo, N., and S. Sato. 2013. Separating information maximum likelihood estimation of realized volatility and covariance with micro-market noise. *North American Journal of Economics and Finance* 26: 282–309.

Kunitomo, N., H. Misaki, and S. Sato. 2015. The SIML estimation of integrated covariance and hedging coefficient under round-off errors, micro-market price adjustments and random samplings. *Asia-Pacific Financial Markets* 22: 333–368.

Misaki, H., and N. Kunitomo. 2015. On robust properties of the SIML estimation of volatility under micro-market noise and random sampling. *International Review of Economics and Finance*.

Miyahara, Y. 2012. *Option pricing in Incomplete Markets: Modeling based on geometric levy processes and minimal entropy martingale measures*. Imperial College Press.

Index

© The Author(s) 2018
N. Kunitomo et al., *Separating Information Maximum Likelihood Method for High-Frequency Financial Data*, JSS Research Series in Statistics, https://doi.org/10.1007/978-4-431-55930-6

Printed in the United States
By Bookmasters